The Future of Tomorrow:

How Technology, Medicine, Computers, and Travel Will Change Beyond the 21st Century

Roderick L. Fennell, M.S. M.I.S.

authorHOUSE®

AuthorHouse™
1663 Liberty Drive, Suite 200
Bloomington, IN 47403
www.authorhouse.com
Phone: 1-800-839-8640

First published by AuthorHouse 9/19/2007

ISBN: 978-1-4343-0293-9 (sc)

Printed in the United States of America
Bloomington, Indiana

This book is printed on acid-free paper.

Acknowledgement...

I would like to first of all give thanks to my Creator, for without Him, none of this would have been possible.

I next would like to thank both of my Parents for always being there for me. For allowing me to fall down and allowing me to learn, through experience, how to pick myself up once more.

This is for the promise I made to my father, Romey, before he passed, who received his Master's Degree from the University of North Texas and hours toward his Doctoral Degree. You were a role model not only to your children but to all of those whose life you touched.

This is for my Mother, Clarice, who received her Master's Degree from Colorado College and hours toward her Doctoral Degree. Thanks for always loving me Mom and for believing in me.

This is also for my Grandmother, Odessa, who received her Master's Degree and was blessed to live on this earth for 95 years. I would also like to acknowledge Vivian and Romie, also my Grandparents that always loved me and tried to teach me right from wrong.

I also acknowledge my brother, Romey III, who is also no longer with us but, always loved his little brother and tried to protect him the best way he knew how.

And finally, I would like to give thanks to my Wife, Dr. Henrilyn for her encouragement and for loving a small town boy as myself. Most of all, I would like to thank her for giving us a wonderful daughter, a product of Italy named Talia E. G. Fennell.

Talia means "a drop of dew from Heaven", which you most certainly are. E. is after her Grandmother, Florence, and G. after her Grandmother Clarice.

To my Father-in-law and Mother-in-law, Florence&Henry, thanks for making me a part of your family and for making such a lovely daughter. Talia, always remember, Daddy loves you and if you believe in Him, all things are possible.....

Table of Contents

PART I

Technological Advances of the 21st Century and Beyond

INTRODUCTION

Mankind has always thought of what if. If I could only find a way to get from Point A to point B faster, I would have more time for X, Y, & Z. From the first invention of the flint rock for hunting, to the discovery of fire, we have always looked for ways to do things just a little better than the day before. That same mind set led man to discover that the world was round instead of flat, that a country called America existed, that machines could work faster than man, i.e. (The Industrial Revolution), and to our technological advances of the 21st century.

Just as society has evolved, so has man and his desires. For those that owned the 1st Horse and buggy, the sight of the Model-T was not only foreign but also laughable. How could something that looked so strange ever catch on? There have been many first from various disciplines that when they initially took place, critics vowed it would never work or catch on but, they were all in time silenced by the masses or its popularity. From the first heart transplant, to the first 2 ton computer of the 50's, mankind continues to kick sand in the face of adversity and find ways to achieve the once impossible.

New technologies surround us, making our lives simpler, more interesting, and more enjoyable. We wear eyeglasses that automatically darken into sunglasses when we step outside on a bright sunny day. Voicemail systems make it easy to leave a message for someone. The Space Invaders and Pac-Man of old can now teach us about music, art, and history. With every breakthrough, a hundred people might say "Wow," but there's always one who says, "How can I push this a little further?" One Inventor's breakthrough is another inventor's stepping stone.

Medicine, biology, quantum physics, computer engineering, chemistry, aerospace, and business are just a few of the areas where you can find current cutting-edge technology (from nanotechnology to neural networks with artificial intelligence). This book examines over a dozen topics on the cutting edge of technology. The affect on our lives and commerce, as well as, where all of this technology is leading in the years to come will be addressed. Just a few years ago, this information might have been well suited for a science-fiction novel, but today it defines what's new on the scientific front.

Roderick L. Fennell, M.S. M.I.S.

Nanotechnology: Putting the Atoms Where We Want Them

Imagine tiny computers so small they must be seen with a microscope. Consider machines smaller than a red blood cell that can circulate through your body and attack and remove infectious organisms. Imagine the creation of ultra light materials that are 10 times as strong as steel or the creation of Molecular computers that could contain storage devices capable of storing trillions of bytes of information in a structure the size of a sugar cube. All of these notions are ideas behind nanotechnology.

Artificial Intelligence: Thinking Machines

What is intelligence? What is cognition? Is thinking and thought uniquely human? Can Artificial Intelligent machines experience emotion and feelings, such as love and anger? A crucial factor in competing in business in the 21st Century will be the clever use of information technology. Today's IT systems are mostly data and communication tools for human workers. Tomorrow's IT systems will be able to do more: automate decisions, intelligently analyze large amounts of data, and learn from their mistakes.

The evolution of Ultra Intelligent (UI), Artificial Intelligence (AI) machines that are themselves a new species may be just a few years away. It is predicted that computers will match the computational functions of the human brain early in the next century, and that soon afterwards humans and computers will merge to become a new species. Can humans compete?

Biotechnology/Genetic Engineering: Cloning, Fighting Disease, Gene Therapy

We have cloned sheep, mice, and cows, so what's to stop scientist from cloning a human? Like computer chips, which perform millions of mathematical operations a second, biochips can perform thousands of biological reactions, such as decoding genes in seconds. Imagine being able to cure cancer by drinking a medicine stirred into your favorite fruit juice.

Widespread use of biochips could remove the guesswork from early treatment of many diseases. Imagine using different DNA-splicing techniques, perhaps within a few years, to attempt to cure two inherited genetic diseases in fetuses. Just place your order for a new born baby with blue eyes, blond hair, higher cognition, or whatever trait a couple desires, and then wait 9 months. It could be only a few years away.

Quantum Teleportation: New York to California in 2 seconds, how about it?

"Beam me up, Scotty," may be here sooner than you think, say scientist from the California Institute of Technology. What is noise now could eventually become messages. Scientist are hoping that quantum computers, which move information about in this way rather than by wires and silicon chips, will be infinitely faster and more powerful than present-day computers.

In the future, teleportation could be used to send information to create replicas of objects, not just beams of light. In 1993, a group of six scientists from IBM confirmed the intuitions of the majority of science fiction writers by showing that perfect teleportation is indeed possible in principle, but only if the original is destroyed. So," Will the real Captain Kirk please stand up! "

Computer Engineering: Distributed Computing, Quantum Computers, Scalable TCP

As microprocessor throughput approaches the speed limitations imposed by fundamental device technology, computational parallelism becomes the most viable alternative for achieving breakthroughs in computing power. Just as hardware advances drive multimedia applications, new multi-media applications, in turn, also increase the appetite for more computational power. No matter how fast computers become, new applications are found that stretch the available resources to the limit. Recent advances in technology are providing faster microprocessors and network communications, reducing power dissipation in electronic systems, and producing higher-density, low-cost data storage devices. In turn, these advances are creating a demand for new multi-media applications and interfaces. Continued advances in computing, communication, and storage technologies, combined with the development of a national and global Grid system, holds the promise of providing the required capacities and an effective environment for computing and science.

PART II

21ST CENTURY TECHNOLOGICAL ADVANCES

1

Nanotechnology: Putting the Atoms Where We Want Them

Imagine a supercomputer no bigger than a human cell. Imagine a four-person, surface-to-orbit spacecraft no larger and no more expensive than the family car. Imagine being able to cure cancer by drinking a medicine stirred into your favorite fruit juice. These are just a few of products expected from Nanotechnology.

If you want to bet on what technology will be important in the future, experts agree that if you choose Nanotechnology, you'll be a big winner. Compared to fractals, virtual reality, and other technologies making news today, Nanotechnology is different in two ways. First, it is a technology just entering development: the first commercial products are still years away. Second, the scale and impact of Nanotechnology will be immense. Rather than being an interesting single technique or application, Nanotechnology will be the basis of humanity's next great technical expansion. MIT's Marvin Minsky, well-known computer scientist and artificial intelligence pioneer, says Nanotechnology could have more effect on our material existence than those last two great inventions in that domain – the replacement of sticks and stones by metals and cement and the harnessing of electricity."[1]

Well, if Nanotechnology is suppose to be so important to our material existence then, let's first start by finding out what it is. Nanotechnology is an umbrella term that covers many areas of research dealing with objects that are measured in nanometers. A nanometer (nm) is a billionth of a meter, or a millionth of a millimeter. Let's now learn how nanomachines will manufacture products, and what will be the impact.

Building with Atoms

Atoms are the building blocks for all matter in our universe. You and everything around you are made of atoms. Nature has perfected the science of manufacturing matter molecularly. For instance, our bodies are assembled in a specific manner from millions of living cells. Cells are nature's nanomachines. Humans still have a lot to learn about the idea of constructing materials on such a small scale. Consumer goods that we buy are made by pushing piles of atoms together in a bulky, imprecise manner. Imagine if we could manipulate each individual atom of an object. That's the basic idea of nanotechnology, and many scientists believe that we are only a few decades away from achieving it.

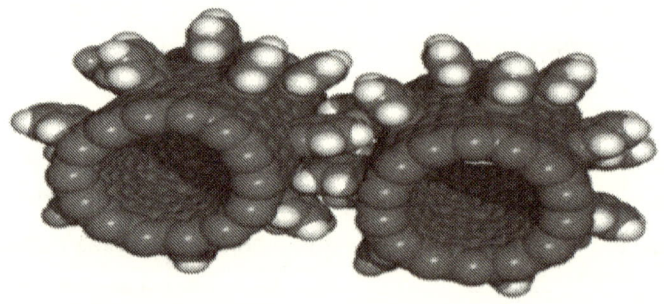

Photo courtesy NASA, Ames Nanogears no more than a nanometer wide could be used to construct a matter compiler, which could be fed raw material to arrange atoms and build a macro-scale structure.

Nanotechnology is a hybrid science combining engineering and chemistry. Atoms and molecules stick together because they have complementary shapes that lock together, or charges that attract. Just like with magnets, a positively charged atom will stick to a negatively charged atom. As millions of these atoms are pieced together by nanomachines, a specific product will begin to take shape. The goal of nanotechnology is to manipulate atoms individually and place them in a pattern to produce a desired structure. There are three steps to achieving nanotechnology-produced goods:

Lets Review:

Scientists must be able to manipulate individual atoms. This means that they will have to develop a technique to grab single atoms and move them to desired positions. In 1990, IBM researchers showed that it is possible to manipulate single atoms. They positioned 35 xenon atoms on the surface of a nickel crystal, using an atomic force microscopy instrument. These positioned atoms spelled out the letters "IBM."

The next step will be to develop nanoscopic machines, called assemblers, that can be programmed to manipulate atoms and molecules at will. It would take thousands of years for a single assembler to produce any kind of material one atom at a time. Trillions of assemblers will be needed to develop products in a viable time frame. In order to create enough assemblers to build consumer goods, some nanomachines, called replicators, will be programmed to build more assemblers.

Trillions of assemblers and replicators will fill an area smaller than a cubic millimeter, and will still be too small for us to see with the naked eye. Assemblers and replicators will work together like hands to automatically construct products, and will eventually replace all traditional labor methods. This will vastly decrease manufacturing costs, thereby making consumer goods plentiful, cheaper and stronger. In the next section, you'll find out how nanotechnology will impact every facet of society, from medicine to computers.[2]

A New Industrial Revolution

In January 2000, U.S. President Bill Clinton requested a $227-million increase in the government's investment in nanotechnology research and development, which included a major initiative called the National Nanotechnology Initiative **(NNI)**. This initiative nearly doubled America's 2000 budget investment in nanotechnology, bringing the total invested in nanotechnology to $497 million for the 2001 national budget. In a written statement, White House officials said that "nanotechnology is the new frontier and its potential impact is compelling."

About 70 percent of the new nanotechnology funding will go to university research efforts, which will help meet the demand for workers with nanoscale science and engineering skills. The initiative will also fund the projects of several governmental agencies, including the National Science Foundation, the Department of Defense, the Department of Energy, the National Institutes of Health, NASA and the National Institute of Standards and Technology. Much of the research will take more than 20 years to complete, but the process itself could touch off a new industrial revolution. Nanotechnology is likely to change the way almost everything, including medicine, computers

11

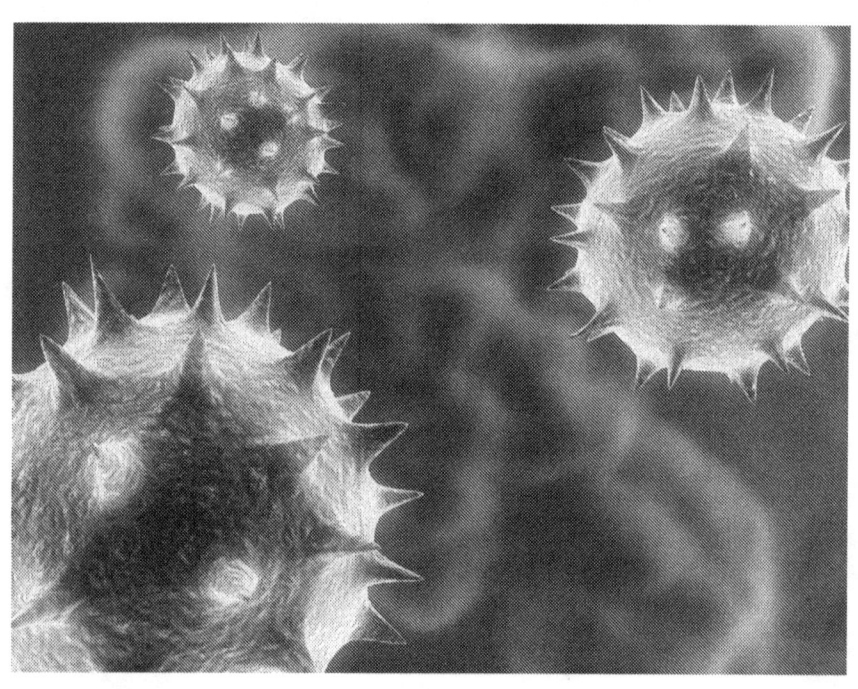

and cars, are designed and constructed. Nanotechnology is anywhere from five to 15 years in the future, and we won't see dramatic changes in our world right away. But let's take a look at the potential effects of nanotechnology:

The first products made from nanomachines will be stronger fibers. Eventually, we will be able to replicate anything, including diamonds, water and food. Famine could be eradicated by machines that fabricate foods to feed the hungry.

In the computer industry, the ability to shrink the size of transistors on silicon microprocessors will soon reach its limits. Nanotechnology will be needed to create a new generation of computer components. Molecular computers could contain storage devices capable of storing trillions of bytes of information in a structure the size of a sugar cube.

Nanotechnology may have its biggest impact on the medical industry. Patients will drink fluids containing nanorobots programmed to attack and reconstruct the molecular structure of cancer cells and viruses to make them harmless. There's even speculation that nanorobots could slow or reverse the aging process, and life expectancy could increase significantly. Nanorobots could also be programmed to perform delicate surgeries -- such nanosurgeons could work at a level a thousand times more precise than the sharpest scalpel. By working on such a small scale, a nanorobot could operate without leaving the scars that conventional surgery does. Additionally, nanorobots could change your physical appearance. They could be programmed to perform cosmetic surgery, rearranging your atoms to change your ears, nose, eye color or any other physical feature you wish to alter.

Nanotechnology has the potential to have a positive effect on the environment. For instance, airborne nanorobots could be programmed to rebuild the thinning ozone layer. Contaminants could be automatically removed from water sources, and oil spills could be cleaned up instantly. Manufacturing materials using the bottom-up method of nanotechnology also creates less pollution than conventional manufacturing processes. Our dependence on non-renewable resources would diminish with nanotechnology. Many resources could be constructed by nanomachines. Cutting down trees,

mining coal or drilling for oil may no longer be necessary. Resources could simply be constructed by nanomachines.[3]

Nanotechnology discussed before the United States Senate

Nanotechnology is also encroaching more and more into the politics of today. U.S. Senator Evan Bayh (D-IN) and U.S. Senator Bill Frist (R-TN) gathered together with the nation's leading experts in the field of nanotechnology, April 5, 2000, to testify before the Senate Science Caucus as it explores the development of this new scientific research with potentially revolutionary applications in the areas of health, commerce and defense.

Bayh stated, "The possibilities of this new technology are limited only by our imagination. It could lead to discoveries that will change the way almost everything - from vaccines to computers - are designed and made, what transistors were to the 1950s, and the Internet and biotechnology are today, nanotechnology may be for the 2010s and 2020s."

The purpose of the panel was to explore the potential of nanotechnology, and focus on ways the federal government can help promote nanotechnology research. Bayh noted the many Indiana universities on the leading edge of nanotechnology research. "The research that our Indiana schools are conducting holds the key to significant economic and social advances. At Purdue University in Lafayette, scientists are building an Ultra-Performance Nanotechnology Center. Potential use of nanoscience to develop next generation computing technologies is one of the main areas of work at Notre Dame University in South Bend," said Bayh as he welcomed Dr. Jim Merz of Notre Dame University as one of the panel witnesses.

Bayh contends that continued scientific innovation is crucial to economic development as well as other possible advances. "Innovation is the key to our comparative advantage in the global economy," Bayh said.

"Federal investment in the physical sciences that help drive innovation - math, chemistry, geology, physics, and chemical, mechanical, and electrical engineering - are all declining, as are the number of college and advanced degrees in these areas," said Bayh. "As a result, the United States risks falling behind other nations such as South Korea, Taiwan, Singapore, Israel, and even Japan. It is vitally important that we continue to support the physical sciences, including nanotechnology, if we are to see increases in productivity and incomes in the years ahead," Bayh said.

The visiting panel of experts at the hearing included: James L. Merz, Ph.D., vice president for Graduate Studies and Research, University of Notre Dame; Don Eigler, Ph.D., IBM fellow, Almaden Research Center, IBM Corporations; Alton D. Romig, Jr. Ph.D., vice president of Science, Technology and Components, and chief technology officer, Sandia National Laboratories; Richard E. Smalley, Ph.D., director, Center for Nanoscale Science and Technology, Rice University.

Bayh and Frist serve on the Senate Science and Technology Caucus, a bipartisan group established to examine the critical role that science and technology play in shaping our modern world. It also provides a venue to discuss federal involvement in science and technology programs. [4]

Nano Tech 2007

The world's 6th International Nanotechnology Exhibition & Conference was held February 21st through 23rd, 2007 at Tokyo Big Sight, Japan. On September 27th through 29th, "Nano Tech 2006" was held in Taiwan. On the local front, the 10th Annual Nanotechnology Conference and Trade Show was also held on May 20th through 24th in Santa Clara, California for all of you that are interested on the National level.

These are some of the topics covered:

Nanotech 2006 Symposium Program

Electronics & Microsystems	Life Sciences & Medicine	Materials & Technologies	Business & Ventures
Nano Electronics & Photonics	Bionano Materials & Tissues	Nano Materials & Devices	Nanotech Business
Nano Fabrication	Bio Sensors & Diagnostics	Soft Nanotech & Applications	Nanotech Ventures
MEMS & NEMS	Biomarkers & Nanoparticles	Polymer Nanotechnology	TechConnect Summit
Sensors & Systems	Cancer Nanotechnology	Carbon Nano Structures & Devices	Nanotech Excellence Awards
Micro & Nano Fluidics	Cellular & Molecular Dynamics	Nano Particles & Applications	NanoSPRINT Virtual Exhibit
MSM - Modeling Microsystems	Computational Modeling	Composites & Interfaces	Expo
WCM - Compact Modeling	Drug Delivery & Therapeutics	Energy Technology & Applications	Nanotech Expo
Homeland Security	Imaging	Environment & Society	BioNano Expo
Controlled	Nano Medicine	ICCN - Nanoscale Modeling	
Clean & Controlled Environments	Nanotech to Neurology	Nanoscale Characterization	CLEAN & CONTROLLED ENVIRONMENTS

They stand confident that nano tech 2006 and 2007 constituted the best national and international conferences in this wide-ranging discipline. There are many issues remaining to be resolved regarding the practical use of nanotechnology, a field of endeavor drawing considerable public attention. But recent technological developments and the awarding of patents are nothing short of amazing and are having a significant impact on researchers at universities and institutes both national and private. Thus nano tech 2007 was able to furnish an excellent opportunity for people to view the technology and resultant products brought about by this fascinating realm of science. Many study meetings and symposia featuring nanotechnology have been held, but none were more exciting than the comprehensive exhibitions to be featured in Japan. Events like these help to create new industries and solidify an industry-university network.[5]

The promises of nanotechnology sound great, don't they? They may even be unbelievable? But, researchers say that we will achieve these capabilities within the next century. And if nanotechnology is, in fact, realized, it might be the human race's greatest scientific achievement yet, completely changing every aspect of the way we live.

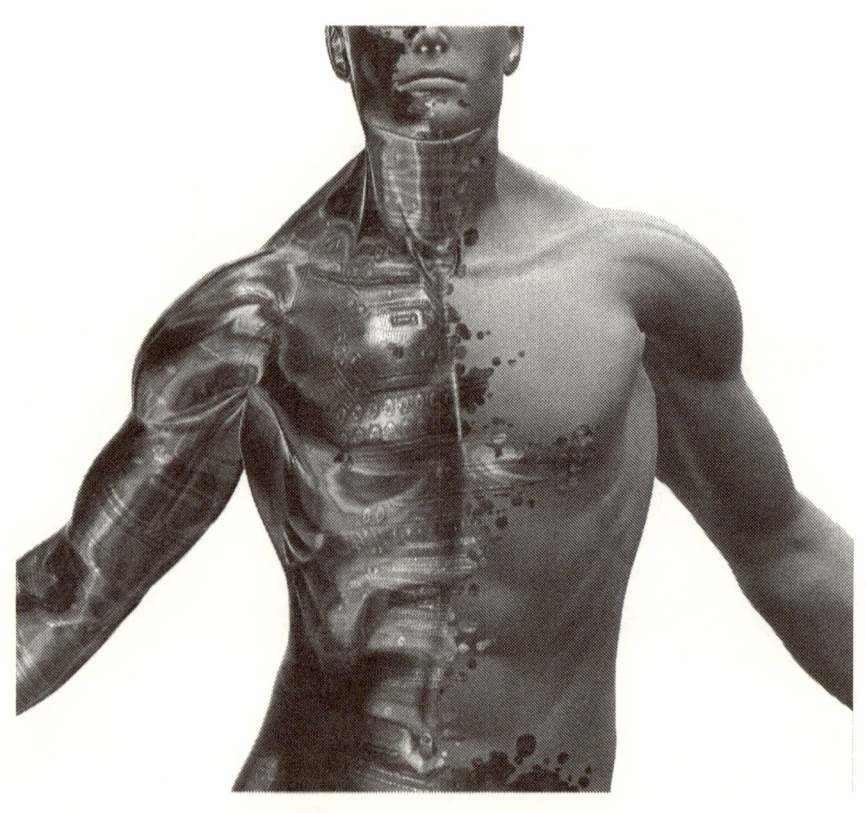

2

ARTIFICIAL INTELLIGENCE: THINKING MACHINES

The Postindustrial society will be fueled not by oil but by a new commodity called Artificial Intelligence (AI). We might regard it as a commodity because it has value and can be traded. Indeed, as will be made clear, the knowledge imbedded in AI software and hardware architectures will become even more salient as a foundation of wealth than the raw materials that fueled the first Industrial Revolution. It is an unusual commodity, because it has no material form. It can be a flow of information with no more physical reality than electrical vibrations in a wire.

If artificial intelligence is the fuel of the second industrial revolution, then we might ask what is it?. There seems to be a difficultly faced when this question is asked. Books on Biology do not generally begin with the question, what is Biology, anyway?

One view is that AI is an attempt to answer a central question that has been debated by scientist, philosophers, and theologians for thousands of years.

How does the human brain – three pounds of "ordinary" matter – give rise to thoughts, feelings, and consciousness? While certainly very complex, our brains are clearly governed by the same physical laws as our machines. [6]

If thought about in this way, the human brain may be regarded as a very capable machine. If given sufficient capacity and the right techniques, our machine may ultimately be able to replicate human intelligence. Being able to finally understand our minds and how they work is truly inspiring. You do not have to fully accept the thought that the human mind is "just" a machine to appreciate both the potential for machines to master many of our intellectual capabilities and the practical implications of doing so.

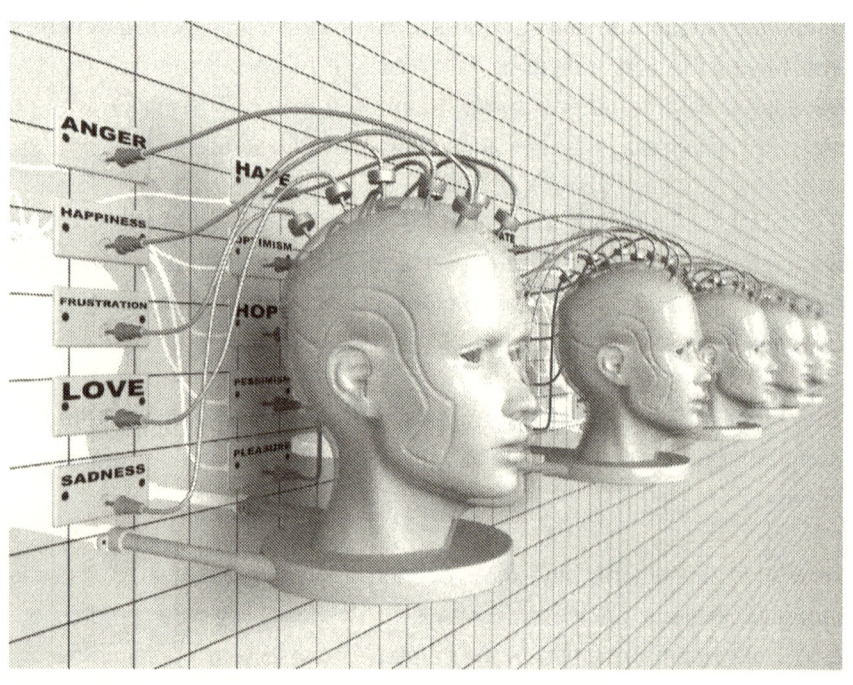

Electronic Circuit Replicates Brain Activity

Scientists have developed an electronic circuit based on operations of the cerebral cortex, the brain's center of intelligence. Although researchers have previously made connections between electronic circuitry and brain circuitry, this is the first artificial circuit to mimic brain activity. The circuit, developed by researchers at the <u>Massachusetts Institute of Technology</u> and at Lucent Technologies' <u>Bell Labs</u>, is made up of artificial neurons that communicate with each other via artificial synapses, in a network modeled after the much larger network of natural neurons in the brain. The artificial network is composed of transistors fabricated on a silicon integrated circuit.

While traditional electronics can be categorized as either analog, like radios, or digital, like computer processors, current research indicates that brain circuitry falls into both categories at once. A good example of the joint analog and digital processes of the brain is everyday perception. If you stand near a fire, your brain receives a great deal of information: You see flickering light, feel heat, smell smoke and detect the movement of the flame. All of this is nonspecific, analog input. But while it is receiving this information, your brain makes a specific digital decision: Is this fire, yes or no?

The new artificial circuit mimics the brain's combination of digital and analog in the synaptic feedback process. In the cerebral cortex, neurons make digital decisions on how to respond to one another based on the collective analog information they receive from the neurons around them.

When the artificial circuit receives analog information from multiple stimuli at the same time, it combines this information to make a decision. For example, when researchers applied simultaneous electrical currents to two artificial neurons, the circuit made a digital decision of which neuron it would respond to and which neuron it would suppress its response to. According to its developers, the circuit does not use any one factor to make its decision; it decides based on the combination of information it receives from all surrounding neurons, in the same way a hungry frog chooses between two flies based on all elements of its perception.

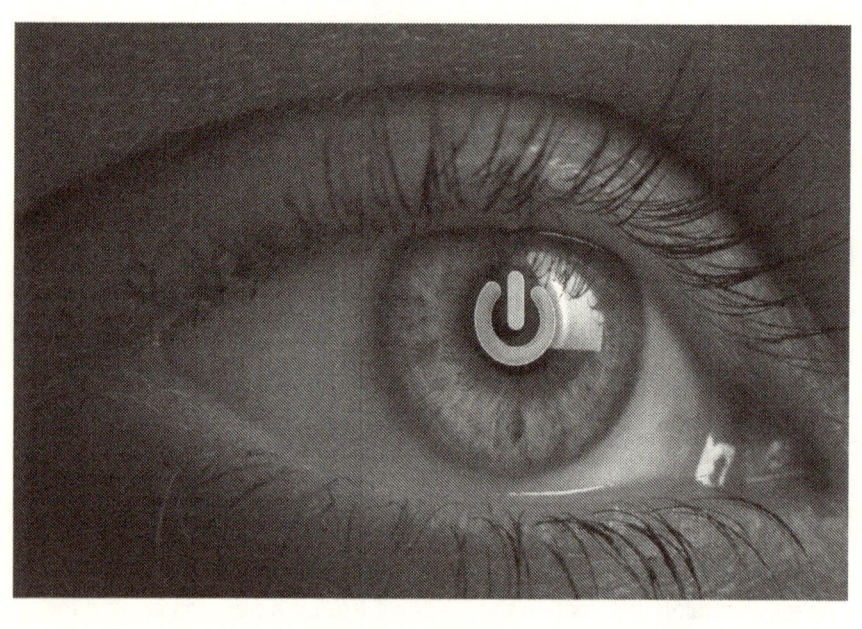

The scientists involved in the project hope the circuit will help advance both electronic circuitry and the study of the brain. They foresee the circuit's unique electronic combination of analog and digital processes eventually leading to machines capable of complex perception, such as recognizing objects on sight. [7]

But computers have a far way to go to match human strengths, and our estimates will depend on analogy and extrapolation. Fortunately, these are grounded in the first bit of the journey, now behind us. Thirty years of computer vision reveals that 1 MIPS can extract simple features from real-time imagery--tracking a white line or a white spot on a mottled background. 10 MIPS can follow complex gray-scale patches--as smart bombs, cruise missiles and early self-driving vans attest. 100 MIPS can follow moderately unpredictable features like roads--as recent long NAVLAB trips demonstrate. 1,000 MIPS will be adequate for coarse-grained three-dimensional spatial awareness--illustrated by several mid-resolution stereoscopic vision programs. 10,000 MIPS can find three-dimensional objects in clutter--suggested by several "bin-picking" and high-resolution stereo-vision demonstrations, which accomplish the task in an hour or so at 10 MIPS. The data fades there--research careers are too short, and computer memories too small, for significantly more elaborate experiments.

There are considerations other than sheer scale. At 1 MIPS the best results come from finely hand-crafted programs that distill sensor data with utmost efficiency. 100-MIPS processes weigh their inputs against a wide range of hypotheses, with many parameters, that learning programs adjust better than the overburdened programmers. Learning of all sorts will be increasingly important as computer power and robot programs grow. This effect is evident in related areas. At the close of the 1980s, as widely available computers reached 10 MIPS, good optical character reading (OCR) programs, able to read most printed and typewritten text, began to appear. They used hand-constructed "feature detectors" for parts of letter shapes, with very little learning. As computer power passed 100 MIPS, trainable OCR programs appeared that could learn unusual typestyles from examples, and the latest and best programs learn their entire data sets. Handwriting recognizers, used by the Post Office to sort mail, and in computers, notably Apple's Newton, have followed a similar path. Speech recognition also fits the model. Under the direction of Raj Reddy, who began his research at Stanford in the 1960s, Carnegie Mellon has led in computer transcription of continuous spoken speech. In 1992 Reddy's group demonstrated a program called Sphinx II on a 15-MIPS workstation with 100 MIPS of specialized signal-processing circuitry. Sphinx II was able to deal with arbitrary English speakers using a several-thousand-word vocabulary. The system's word detectors, encoded in statistical structures known as Markov tables, were shaped by an automatic learning process that digested hundreds of hours of spoken examples from thousands of Carnegie Mellon volunteers enticed by rewards of pizza and ice cream. Several practical voice-control and dictation systems are sold for personal computers today, and some heavy users are substituting larynx for wrist damage.

More computer power is needed to reach human performance, but how much? Human and animal brain sizes imply an answer, if we can relate nerve volume to computation. Structurally and functionally, one of the best understood neural assemblies is the retina of the vertebrate eye. Happily, similar operations have been developed for robot vision, handing us a rough conversion factor.

It takes robot vision programs about 100 computer instructions to derive single edge or motion detections from comparable video images. 100 million instructions are needed to do a million detections, and 1,000 MIPS to repeat them ten times per second to match the retina.

The 1,500 cubic centimeter human brain is about 100,000 times as large as the retina, suggesting that matching overall human behavior will take about 100 million MIPS of computer power. The most powerful experimental supercomputers in 1998, composed of thousands or tens of thousands of the fastest microprocessors and costing tens of millions of dollars, can do a few million MIPS. They are within striking distance of being powerful enough to match human brainpower, but are unlikely to be applied to that end. Why tie up a rare twenty-million-dollar asset to develop one ersatz-human, when millions of inexpensive original-model humans are available? Such machines are needed for high-value scientific calculations, mostly physical simulations, having no cheaper substitutes. AI research must wait for the power to become more affordable.

If 100 million MIPS could do the job of the human brain's 100 billion neurons, then one neuron is worth about 1/1,000 MIPS, i.e., 1,000 instructions per second. That's probably not enough to simulate an actual neuron, which can produce 1,000 finely timed pulses per second. Our estimate is for very efficient programs that imitate the aggregate function of thousand-neuron assemblies. Almost all nervous systems contain subassemblies that big.[8]

Strong/ Weak Truly Intelligent Machines...

The term Artificial Intelligence was first coined by John McCarthy in 1956 when he proposed that "intelligence can in principle be so precisely described that a machine can be made to simulate it." He now defines A.I. as "the science and engineering of making intelligent machines, especially intelligent computer programs." A.I. is generally associated with Computer Science, but it has many important links with other fields such as Math, Psychology, Cognition, Biology and Philosophy, among many others.

Artificial Intelligence is often divided into two classes: Strong A.I. and Weak A.I... Strong A.I. makes the bold claim that computers can be made to think on a level at least equal to humans; that they are capable of cognitive mental states. This is the kind of A.I. that is portrayed in movies like *Blade Runner* and more recently *A.I.*.. Weak A.I. simply states that some "thinking-like" features can be added to computers to make them more useful tools; that machines can simulate human cognition, in other words act as if they are intelligent. This has already started to happen, for example, speech recognition software.

The Turing Test

The 'Turing Test' is an experiment suggested by mathematician Alan Turing in his 1950 paper *Computing Machinery and Intelligence.* He argued that if a machine could successfully pretend to be human to a knowledgeable observer, then you certainly should consider it intelligent. In the Turing test, a judge has conversations via teletype,

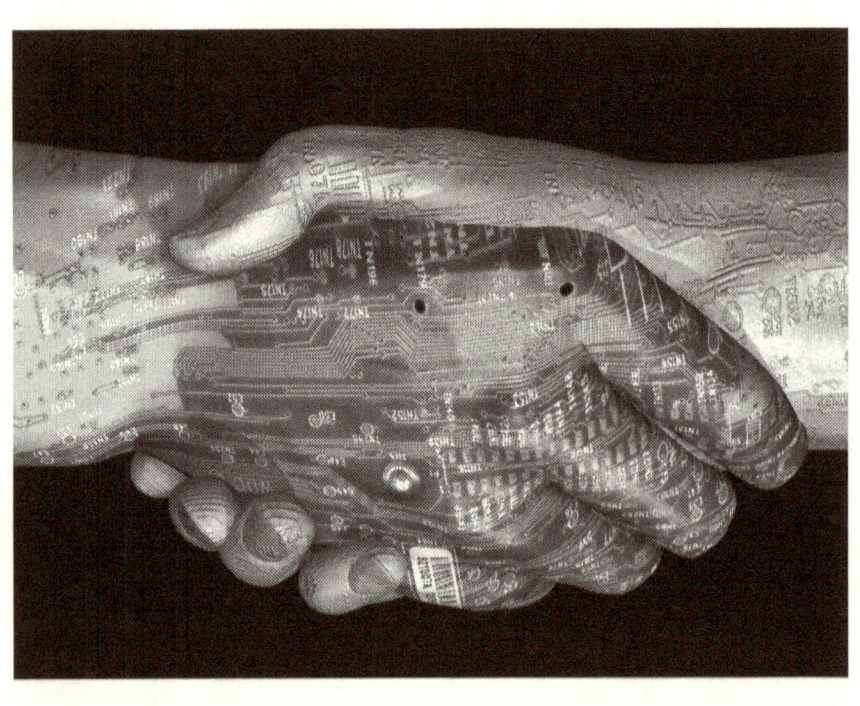

with two systems, one human, the other a machine. The conversations can be about anything, and proceed for a set period of time. If, at the end of this time, the judge cannot distinguish the machine from the human on the basis of the conversation, then Turing argued that we would have to say that the machine was intelligent.

Branches of A.I.

There are many branches of Artificial Intelligence including:

- **Neural Networks** - These are systems that attempt to simulate intelligence by reproducing the types of physical connections that occur in animal brains.

- **Natural Language Processing** - This involves programming computers to understand natural human languages.

- **Robotics** - This field attempts to help robots to act intelligently. For example to see, hear, and react to other sensory stimuli.

- **Game Playing** -This involves programming computers to play games such as chess.

- **Expert systems** - This is where computers are programmed to make decisions in real-life situations. [9]

With the views of Raj Reddy and McCarthy fresh in our minds, let us now expand our thoughts just a little more. In an article by Bill Joy titled, "Why the future doesn't need us," he speaks directly to famed author Ray Kurzweil as to Robotics, genetic engineering, and nanotechnology. The questioned posed was, "are these technologies threatening to make humans an endangered species?"

Bill Joy states," Ray gave me a partial preprint of his then-forthcoming book, **The Age of Spiritual Machines**, which outlined a utopia he foresaw - one in which humans gained near immortality by becoming one with robotic technology. On reading it, my sense of unease only intensified; I felt sure he had to be understating the dangers, understating the probability of a bad outcome along this path.

I found myself most troubled by a passage detailing adystopian scenario:

The New Luddite Challenge

First let us postulate that the computer scientists succeed in developing intelligent machines that can do all things better than human beings can do them. In that case presumably all work will be done by vast, highly organized systems of machines and no human effort will be necessary. Either of two cases might occur. The machines might be permitted to make all of their own decisions without human oversight, or else human control over the machines might be retained.

If the machines are permitted to make all their own decisions, we can't make any conjectures as to the results, because it is impossible to guess how such machines might behave. We only point out that the fate of the human race would be at the mercy of the machines. It might be argued that the human race would never be foolish enough to hand over all the power to the machines. But we are suggesting neither that the human race would voluntarily turn power over to the machines nor that the machines would willfully seize power. What we do suggest is that the human race might easily permit itself to drift into a position of such dependence on the machines that it would have no practical choice but to accept all of the machines' decisions. As society and the problems that face it become more and more complex and machines become more and more intelligent, people will let machines make more of their decisions for them, simply because machine-made decisions will bring better results than man-made ones. Eventually a stage may be reached at which the decisions necessary to keep the system running will be so complex that human beings will be incapable of making them intelligently. At that stage the machines will be in effective control. People won't be able to just turn the machines off, because they will be so dependent on them that turning them off would amount to suicide.

On the other hand it is possible that human control over the machines may be retained. In that case the average man may have control over certain private machines of his own, such as his car or his personal computer, but control over large systems of machines will be in the hands of a tiny elite

- just as it is today, but with two differences. Due to improved techniques the elite will have greater control over the masses; and because human work will no longer be necessary the masses will be superfluous, a useless burden on the system. If the elite are ruthless they may simply decide to exterminate the mass of humanity. If they are humane they may use propaganda or other psychological or biological technique to reduce the birth rate until the mass of humanity becomes extinct, leaving the world to the elite. Or, if the elite consist of soft-hearted liberals, they may decide to play the role of good shepherds to the rest of the human race. They will see to it that everyone's physical needs are satisfied, that all children are raised under psychologically hygienic conditions, that everyone has a wholesome hobby to keep him busy, and that anyone who may become dissatisfied undergoes "treatment" to cure his "problem." Of course, life will be so purposeless that people will have to be biologically or psychologically engineered either to remove their need for the power process or make them "sublimate" their drive for power into some harmless hobby. These engineered human beings may be happy in such a society, but they will most certainly not be free. They will have been reduced to the status of domestic animals.[10]

Just to validate the author's position more, Bill Joy is the cofounder and Chief Scientist of Sun Microsystems, was co chair of the presidential commission on the future of IT research, and is coauthor of, The Java Language Specification. His work on the Jini pervasive computing technology was featured in Wired 6.08. Fellow Bill Joy's (Sun Microsystems) article in *Wired* this spring expanded on a topic he addressed at his 1999 induction into the Academy: "How to respond to the risks and benefits of rapid innovation in robotics, nanotechnology, and biotechnology." With his help, the Academy convened three meetings during the spring and summer of 2000 to reflect on the *Social Implications of the New Technologies*. More than thirty Fellows and other scholars -- specialists in molecular biology, computational and information sciences, business, and the humanities -- participated in the discussion of how society can best evaluate the risks and benefits of revolutionary advances in these 21st-century technologies. In his address to the Academy, Joy spoke as a representative of the newly inducted members in the mathematical and physical sciences (Class I). He warned that the development of powerful computers, a million times more powerful than today's personal computers, coupled with the technology to catalogue human genes and to construct material at the atomic level, will allow us to determine the fate of our species. "Science is providing possibilities but no useful limits," Joy said, so "our choices should come from spiritual, artistic, and ethical values." Several participants in the Academy seminars questioned Joy's assertion that current risks were greater than those associated with earlier technologies such as nuclear arms and genetic engineering. They felt that the Academy should also address growing public misinformation about the benefits of the new technologies and reflected on what lessons for the future could be drawn from previous efforts to grapple with the consequences of technological advances. A volume to explore risks and benefits from an ethical, historical, and technological perspective is now being planned.[11]

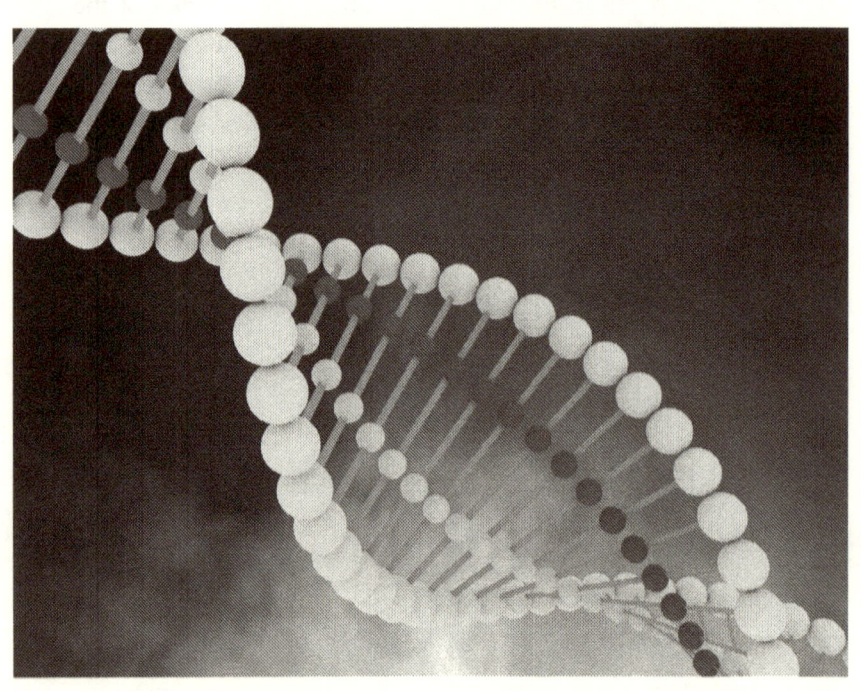

3

Biotechnology/Genetic Engineering: DNA Computers, Cloning, Genetics in the Womb

Even as you read this page, computer chip manufacturers are furiously racing to make the next <u>microprocessor</u> that will topple speed records. Sooner or later, though, this competition is bound to hit a wall. Microprocessors made of silicon will eventually reach their limits of speed and miniaturization. Chip makers need a new material to produce faster computing speeds.

You won't believe where scientists have found the new material they need to build the next generation of microprocessors. Millions of natural supercomputers exist inside living organisms, including your body. DNA (deoxyribonucleic acid) molecules, the material our genes are made of, have the potential to perform calculations many times faster than the world's most powerful human-built computers. DNA might one day be integrated into a computer chip to create a so-called biochip that will push computers even faster. DNA molecules have already been harnessed to perform complex mathematical problems.

While still in their infancy, DNA computers will be capable of storing billions of times more data than your personal computer. In this section 3, you will learn how scientists are using genetic material to create nano-computers that might take the place of silicon-based computers in the next decade.

DNA computers can't be found at your local electronics store yet. The technology is still in development, and didn't even exist as a concept a decade ago. In 1994, Leonard Adleman introduced the idea of using DNA to solve complex mathematical problems. Adleman, a computer scientist at the University of Southern California, came to the conclusion that DNA had computational potential after reading the book "Molecular Biology of the Gene," written by James Watson, who co-discovered the structure of DNA in 1953. In fact, DNA is very similar to a computer hard drive in how it stores permanent information about your genes.

Adleman is often called the inventor of DNA computers. His article in a 1994 issue of the journal Science outlined how to use DNA to solve a well-known mathematical problem, called the directed Hamilton Path problem, also known as the "traveling salesman" problem. The goal of the problem is to find the shortest route between a number of cities, going through each city only once. As you add more cities to the problem, the problem becomes more difficult. Adleman chose to find the shortest route between seven cities.

You could probably draw this problem out on paper and come to a solution faster than Adleman did using his DNA test-tube computer. Here are the steps taken in the Adleman DNA computer experiment:

Strands of DNA represent the seven cities. In genes, genetic coding is represented by the letters A, T, C and G. Some sequence of these four letters represented each city and possible flight path.

These molecules are then mixed in a test tube, with some of these DNA strands sticking together. A chain of these strands represents a possible answer.

Within a few seconds, all of the possible combinations of DNA strands, which represent answers, are created in the test tube. Adleman eliminates the wrong molecules through chemical reactions, which leaves behind only the flight paths that connect all seven cities.

The success of the Adleman DNA computer proves that DNA can be used to calculate complex mathematical problems. However, this early DNA computer is far from challenging silicon-based computers in terms of speed. The Adleman DNA computer created a group of possible answers very quickly, but it took days for Adleman to narrow down the possibilities. Another drawback of his DNA computer is that it requires human assistance. The goal of the DNA computing field is to create a device that can work independent of human involvement.

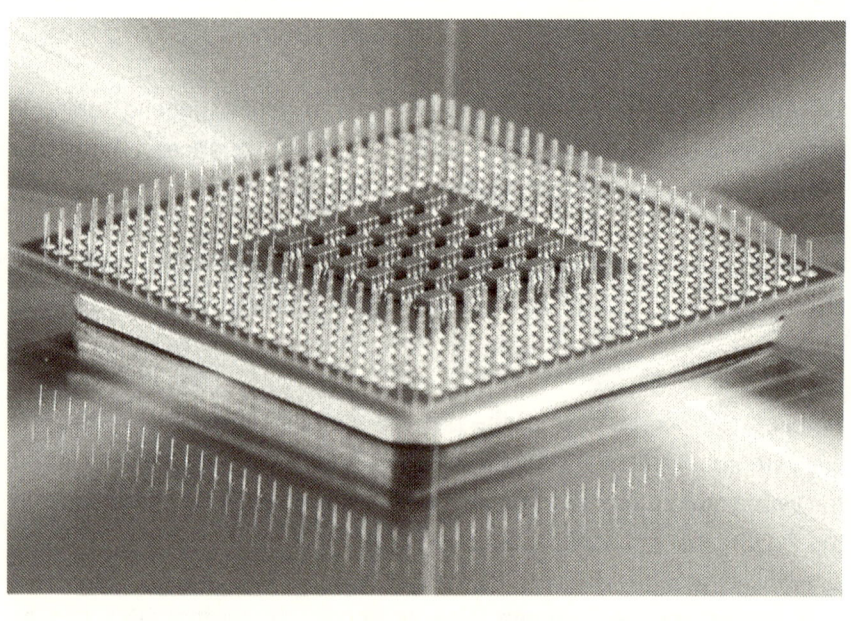

Three years after Adleman's experiment, researchers at the University of Rochester developed logic gates made of DNA. Logic gates are a vital part of how your computer carries out functions that you command it to do. These gates convert binary code moving through the computer into a series of signals that the computer uses to perform operations. Currently, logic gates interpret input signals from silicon transistors, and convert those signals into an output signal that allows the computer to perform complex functions.

The Rochester team's DNA logic gates are the first step toward creating a computer that has a structure similar to that of an electronic PC. Instead of using electrical signals to perform logical operations, these DNA logic gates rely on DNA code. They detect fragments of genetic material as input, splice together these fragments and form a single output. For instance, a genetic gate called the "And gate" links two DNA inputs by chemically binding them so they're locked in an end-to-end structure, similar to the way two Legos might be fastened by a third Lego between them. The researchers believe that these logic gates might be combined with DNA microchips to create a breakthrough in DNA computing.

DNA computer components—logic gates and biochips—will take years to develop into a practical, workable DNA computer. If such a computer is ever built, scientists say that it will be more compact, accurate and efficient than conventional computers. [12]

A Successor to Silicon

Silicon microprocessors have been the heart of the computing world for more than 40 years. In that time, manufacturers have crammed more and more electronic devices onto their microprocessors. In accordance with Moore's Law, the number of electronic devices put on a microprocessor has doubled every 18 months. **Moore's Law** is named after Intel founder Gordon Moore, who predicted in 1965 that microprocessors would double in complexity every two years. Many have predicted that Moore's Law will soon reach its end, because of the physical speed and miniaturization limitations of silicon microprocessors.

D Moore's Law leaves off that there are several advantages to using DNA instead of silicon: As long as there are cellular organisms, there will always be a supply of DNA.

The large supply of DNA makes it a cheap resource. Unlike the toxic materials used to make traditional microprocessors, DNA **biochips** can be made cleanly. DNA computers are many times smaller than today's computers.

DNA's key advantage is that it will make computers smaller than any computer that has come before them, while at the same time holding more data. One pound of DNA has the capacity to store more information than all the electronic computers ever built; and the computing power of a teardrop-sized DNA computer, using the DNA logic gates, will be more powerful than the world's most powerful supercomputer. More than 10 trillion DNA molecules can fit into an area no larger than 1 cubic centimeter (0.06 cubic inches). With this small amount of DNA, a computer would be able to hold 10 terabytes of data, and perform 10 trillion calculations at a time. By adding more DNA, more calculations could be performed.

Unlike conventional computers, DNA computers perform calculations parallel to other calculations. Conventional computers operate linearly, taking on tasks one at a time. It is parallel computing that allows DNA to solve complex mathematical problems in hours, whereas it might take electrical computers hundreds of years to complete them.

The first DNA computers are unlikely to feature word processing, e-mailing and solitaire programs. Instead, their powerful computing power will be used by national governments for cracking secret codes, or by airlines wanting to map more efficient routes. Studying DNA computers may also lead us to a better understanding of a more complex computer—the human brain. [13]

Cloning

DNA logic gates are but just one way technology is progressing in the 21st century. Nothing really prepared the world for the February 23, 1997

announcement that Ian Wilmut, a Scottish scientist, and his colleagues at the Roslin Institute successfully used a technique called somatic cell nuclear transfer (SCNT) to create a clone of a sheep; the cloned sheep was named Dolly.

SCNT involves transferring the nucleus of an adult sheep somatic cell, into a sheep egg from which the nucleus had been removed. After nearly 300 attempts, the cloned sheep known as Dolly was born to a surrogate sheep mother.

There's little doubt that within the next decade, we will hear a more shocking announcement of the first cloned human. Several groups have developed plans to be the first to do so, and the research is already underway to make it happen.

SCNT is not reproduction since a sperm cannot be used with the technique, but rather it is an extension of technology used not only in research but also used to produce medically relevant cellular products such as cartilage cells for knees, as well as gene therapy products. On February 28, 1997, FDA announced a comprehensive plan for the regulation of cell and tissue based therapies that incorporated the legal authorities described in FDA's 1993 guidance "Proposed Approach to Regulation of Cellular and Tissue-Based Products ." [14]

On March 7, 1997 then President Clinton issued a memorandum that stated: "Recent accounts of advances in cloning technology, including the first successful cloning of an adult sheep, raise important questions. They potentially represent enormous scientific breakthroughs that could offer benefits in such areas as medicine and agriculture. But the new technology also raises profound ethical issues, particularly with respect to its possible use to clone humans." (Prohibitions on Federal Funding for Cloning of Human Beings).

The memorandum explicitly prohibited Federal Funding for cloning of a human being, and also directed the National Bioethics Advisory Commission (NBAC) to thoroughly review the legal and ethical issues associated with the use of cloning technology to create a human being. "NBAC found that concerns relating to the potential psychological harms to children and effects on the moral, religious, and cultural values of society merited further reflection and deliberation." The report, Ethical Issues in Human Stem Cell Research, September 1999, describes 5 recommendations.

Somatic cell nuclear transfer holds great potential to someday create medically useful therapeutic products. FDA believes, however, that there are major unresolved questions pertaining to the use of cloning technology to clone a human being which must be seriously considered and resolved before the Agency would permit such investigation to proceed. The Agency sent a "Dear Colleague" letter which stated that creating a human being using cloning technology is subject to FDA regulation under the Public Health Service Act and the Food Drug and Cosmetic Act. This letter notified researchers that clinical research using SCNT to create a human being could precede only when an investigational new drug application (IND) is in effect. Sponsors are required to submit to FDA.

Recently, FDA sent letters to remind the research community that FDA jurisdiction over clinical research using cloning technology to create a human being, and to advise that FDA regulatory process is required in order to initial these investigations.

On March 28, 2001, Dr. Kathryn C. Zoon, Director, Center for Biologics Evaluation and Research gave testimony before the Subcommittee on Oversight and Investigations Committee on Energy and Commerce, United States House of Representatives. Her statement described FDA's role in regulating the use of cloning technology to clone a human being and further described current significant scientific concerns in this area.[15]

In January 2001, a small consortium of scientists led by Panayiotis Zavos, a former University of Kentucky professor, and Italian researcher Severino Antinori said that they planned **to clone a human in the next two years**. At about the same time, the New York Post reported a story about an American couple who planned to pay $500,000 to Las Vegas-based Clonaid for a clone of their deceased infant daughter.

These scientists may be chasing glory in the name of science. Whatever their motivation, it's likely that we will see the first cloned human baby appear on the evening news perhaps as soon as **2007.** Scientists have shown that current cloning techniques work, but only rarely do they succeed in creating a cloned embryo that makes it through birth.

If human cloning proceeds, scientists plan to use somatic cell nuclear transfer, which is the same procedure that was used to create Dolly the sheep. Somatic cell nuclear transfer begins when doctors take the egg from a donor and remove the nucleus of the egg, creating an enucleated egg. A cell, which contains DNA, is then taken from the person who is being cloned. The enucleated egg is then fused together with the cloning subject's cell using electricity. This creates an embryo, which is implanted into a surrogate mother through in vitro fertilization. If the procedure is successful, then the surrogate mother will give birth to a baby that is a clone of the cloning subject at the end of a normal gestation period. Of course, the success rate is only about one or two out of 100 embryos. It took 300 attempts to create Dolly. Take a look at the graphic to see how the somatic cell nuclear transfer cloning process works.

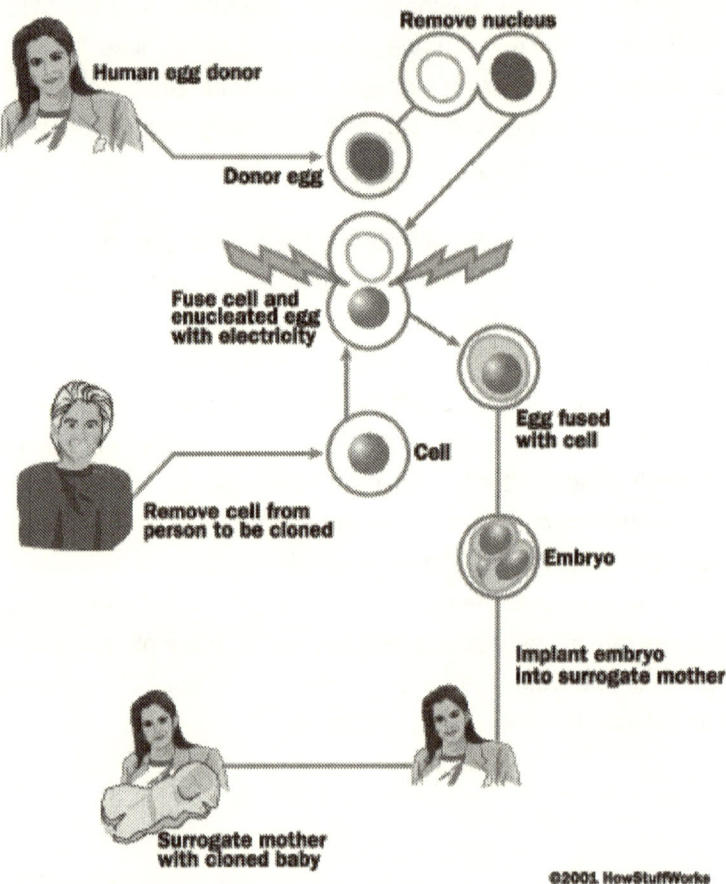

Some scientists seem to think that human cloning is inevitable, but why would we want to clone people? There are many reasons that would make people turn to cloning. Let's explore a few of these reasons.

Who Will Clone?

Not all cloning would involve creating an entirely new human being. Cloning is seen as a possible way to aid some people who have severe medical problems. One potential use of cloning technology would involve creating a human repair kit. In other words, scientists could clone our cells and fix mutated genes that cause diseases. In January 2001, the British government passed rules to allow cloning of human embryos to combat diseases such as Parkinson's and Alzheimer's.

While it may take time for cloning to be fully accepted, therapeutic cloning will likely be the first step in that direction. Therapeutic cloning is the process by which a person's DNA is used to grow an embryonic clone. However, instead of inserting this embryo into a surrogate mother, its cells are used to grow stem cells. These stem cells can be used as a human repair kit. They can grow replacement organs, such as hearts, livers and skin. They can also be used to grow neurons to cure those who suffer from Alzheimer's,

Parkinson's or Rett Syndrome.

Here's how therapeutic cloning works:

DNA is extracted from a sick person.

The DNA is then inserted into an enucleated donor egg.

The egg then divides like a typical fertilized egg and forms an embryo.

Stem cells are removed from the embryo.

Any kind of tissue or organ can be grown from these stem cells to treat the sick.

Others see cloning as a way to aid couples with infertility problems, but who want a child with at least one of the parent's biological attributes. Zavos and Antinori say that helping these couples is the goal of their research. Zavos said that there are hundreds of couples already lined to to pay approximately $50,000 for the service. The group said that the procedure would involve injecting cells from an infertile male into an egg, which would be inserted into the female's uterus. Their child would look the same as the father.

Another use for human cloning could be to bring deceased relatives back to life. Imagine using a piece of your great-grandmother's DNA to create a clone of her. In a sense, you could be the parent of your great-grandmother. This opens the door to many ethical problems, but it's a door that could soon be opened. One American couple, who has had difficulty dealing with the death of their infant daughter, is paying $500,000 to Clonaid to clone their daughter using preserved skin cells.

To Clone or Not to Clone

Critics of cloning repeat the question often associated with controversial science: "Just because we can, does it mean we should?" The closer we come to being able to clone a human, the hotter the debate over it grows. For all the good things cloning may accomplish, opponents say that it will do just as much harm. Another question is how to regulate cloning procedures.

There is no federal law banning cloning in the United States, but several states have passed their own laws to ban the practice. The U.S. Food and Drug Administration (FDA) have also said that anyone in the United States attempting human cloning must first get its permission. In Japan, human cloning is a crime that is punishable by up to 10 years in prison. England has allowed cloning human embryos, but is working to pass legislation to stop total human cloning.

While laws are one deterrent to pursuing human cloning at this time, some scientists believe the technology is not ready to be tested on humans. Ian Wilmut, one of co-creators of Dolly, has even said that human cloning projects would be criminally irresponsible. Cloning technology is still in its early stages, and nearly 98 percent of cloning efforts end in failure. The embryos are either not suitable for implanting into the uterus or they die sometime during gestation or shortly after birth. Those clones that do survive, wind up suffering from fatal or problematic genetic abnormalities. Some clones have been born with defective hearts, lung problems, diabetes, blood vessel problems and malfunctioning immune systems. One of the more famous cases was a cloned sheep that was born but suffered from chronic hyperventilation caused by malformed arteries leading to the lungs. Opponents of cloning will point out that we can euthanize these defective clones of other animals, but they ask what happens if a human clone is born with these same problems. Advocates of cloning, respond that it is now easier to pick out defective embryos even before they are implanted into the mother. The debate over human cloning is just beginning, but as science advances, it could be the biggest ethical dilemma of the 21st century. [16]

The Technology of Gene Therapy

Scientists have confirmed that the first genetically altered humans have been born and are healthy. Up to 30 such children have been born - 15 of them as a result of one experimental programme at a US laboratory.

But the technique has been criticized as unethical by some scientists and would be illegal in many countries, including the United Kingdom. Genetic fingerprint tests on two one-year-old children confirm that they contain a small quantity of additional genes not inherited from either parent. The additional genes were taken from a healthy donor and used to overcome the infertility problems of the mother.

Germ line modification

The additional genes that the children carry have altered their germ line, or their collection of genes that they will pass on to their offspring. Altering the germ line is something that the vast majority of scientists deem unethical given the limitations of our knowledge. It is illegal to do so in many countries and the US Government will not provide funds for any experiment that intentionally or unintentionally alters inherited genes. The children were born following a technique called ooplasmic transfer. This involves taking some of the contents of a donor cell and injecting it into the egg cell of a woman with infertility problems.

The researchers, at the Institute for Reproductive Medicine and Science of St Barnabas in New Jersey, US, believed that some women were infertile because of defects in their mitochondria.

These are tiny structures containing genes that float around inside the cell away from the cell's nucleus, where the vast majority of the genes reside. There can be as many as 100,000 of them floating in the cells cytoplasm.

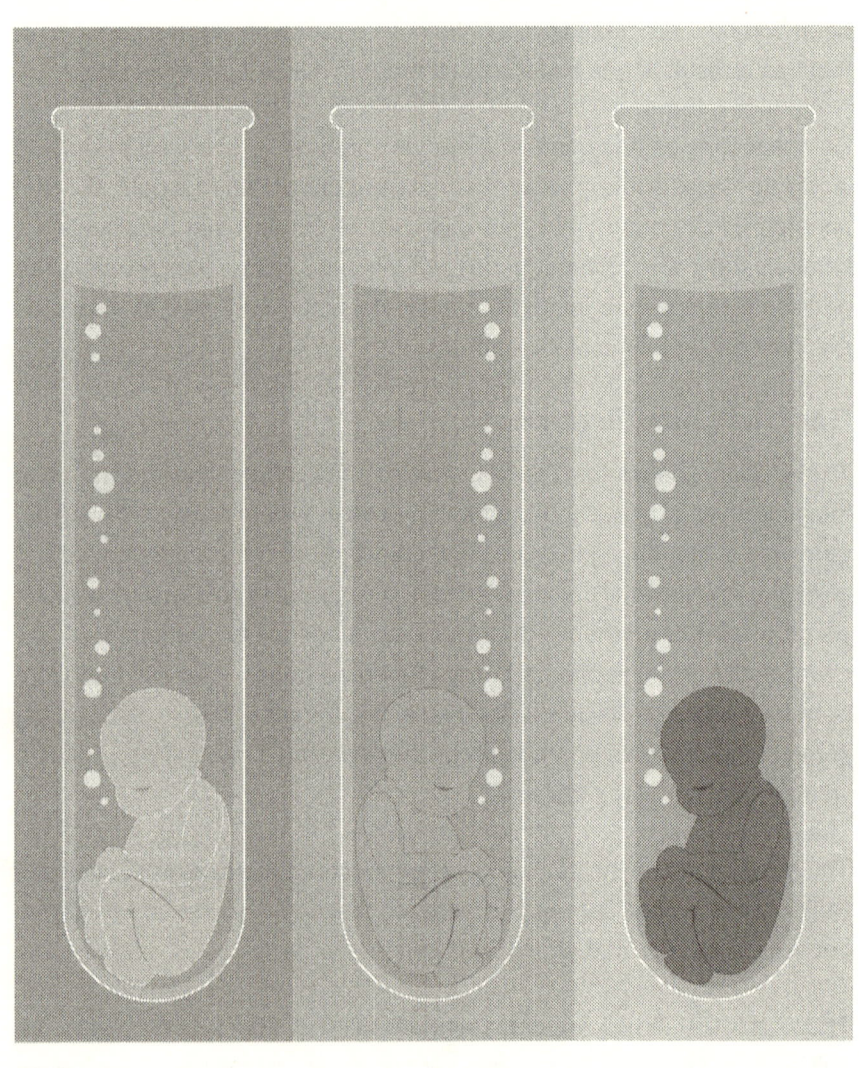

Two mothers

They are essential to cellular energy production and scientists suspect they have many other important, but as yet unappreciated, roles.

Mitochondrial DNA is passed down from generation to generation along the maternal line. The US researchers wanted to supplement a woman's defective mitochondria with healthy ones from a donor.

Having just tested the children born as a result of this procedure, the scientists have confirmed that the children's cells contain mitochondria, and hence genes, from two women as well as their fathers.

Writing in the journal Human Reproduction, the researchers say that this "is the first case of human germ line genetic modification resulting in normal healthy children".

'Great reservations'

British experts have severely criticized the development.

Infertility pioneer Lord Winston of the Hammersmith Hospital in London told BBC News Online that he had great reservations about it. "Regarding the treatment of the infertile, there is no evidence that this technique is worth doing," he said. "I am very surprised that it was even carried out at this stage. It would certainly not be allowed in Britain. "There is no evidence that this is a possible valuable treatment for infertility," he added.

Lord Winston said that, although the number of additional genes involved was tiny, it was in principle the wrong thing to do. The Human Fertilization and Embryology Authority (HFEA), the body that monitors and regulates UK reproductive medical activities, told BBC News Online that it was aware of the technique but had decided not to allow it in the UK because of its uncertainties and the possible alteration of the human germ line. [17]

Gene therapy corrects sickle cell anemia in mice

Roderick L. Fennell, M.S. M.I.S.

Dec 14 (HeartCenterOnline) - Researchers at Harvard Medical School and the Massachusetts Institute of Technology were able to correct sickle cell anemia in mice, according to an article in the December 14 issue of Science. In a press release from the National Heart, Lung and Blood Institute (NHLBI), director Dr. Claude Lenfant hailed the successful experiment as "a significant achievement that brings us closer to human gene therapy for what is a very serious genetic blood disorder."

According to the American Sickle Cell Anemia Association (ASCAA), sickle cell anemia is a condition in which some red blood cells are shaped like half-moons or "sickles" that don't flow easily through the blood vessels like normal, more rounded cells do. As a result, the sickle cells can get caught in smaller blood vessels and obstruct blood flow to vital organs. Symptoms vary, but both children and adults could experience severe episodes of pain or "crises" 15 times a year or more.

To perform the *Science* experiment, the researchers removed bone marrow from mice with the disease and added an anti-sickling gene to it. Then they injected the marrow into a different group of diseased mice, which had been treated with radiation to remove some of their original bone marrow.

Results showed that the treated mice continued to show anti-sickling gene expression in up to 52 percent of their total hemoglobin. In other words, fewer abnormal hemoglobin molecules were present, which means fewer irreversibly sickle cells. One mouse had no detectable irreversibly sickle cells at all. Other problems corrected by the treatment included enlarged spleen and abnormal urine concentration. Researchers believe that humans would benefit from any gene expression greater than 15 percent.

The ASCAA stated, "The ultimate cure for sickle cell anemia may be gene therapy" and noted two different techniques that are currently being investigated. [18]

Science is exploring many applications of gene technology.

I have included prepared tables outlining some of these, which you can view below:

MEDICINE

WHAT	WHY	IN USE IN AUSTRALIA
New drugs	• To design more effective therapies for diseases like cancer, diabetes, influenza and hepatitis C.	yes — Relenza for treatment of influenza. Other drugs are still the subject of research.
Gene therapy	• To replace an abnormal gene with a normal gene.	no
	• To insert a missing gene.	no
	• To switch off rogue genes that may cause cancers.	no
	• To stop viruses replicating within cells.	no
Insulin	• To replace production of insulin from pigs' livers. This has increased efficiency of insulin manufacture while reducing cost.	yes
Therapeutic cloning	• Embryos are cloned for their cells, which can be used in gene therapies. Like all cloning, this technique has significant moral and ethical implications and remains the subject of research.	no
Antibody treatments	• To develop antibodies that will help diagnose and treat diseases like cancer and diabetes.	no

Antibiotic-resistance	• Animal and human health — to find out if and how bacteria from different animals swap genes for resisting antibiotics so as to improve livestock health practices and preserve precious antibiotics.	n/a

ANIMALS (see also article on <u>Biotechnology in the Livestock Industry</u>)

WHAT	WHY	IN USE IN AUSTRALIA
Poultry	• Poultry health — diagnostic kits for infectious bursal disease.	no
Pigs	• Pig health — to boost pigs' natural immunity to infection and so reduce use of antibiotic therapy.	no
Salmon	• Aquaculture — to increase growth rate and size of salmon.	no
Dairy cattle	• Milk — to remove lactose from milk so that people with lactose-intolerance can eat dairy products.	no
Fish and crustaceans	○ Aquaculture — to boost productivity by: increasing growth; and health through inbuilt disease resistance.	no

Advances in technology have led to many great discoveries and opened a vast array of possibilities for mankind. The wonders of Biochips, Cloning, and Gene therapy, are opening up an entirely new set of issues in the 21st century. Beside from the moral implications, we as a species may be able to not only improve the world around us but, in essence improve ourselves.

With increased lifespan, the end of hunger worldwide, and the creation of super humans from our genetic play, what other problems will we face? Will we still be using the same method of transportation also? In our next section, we will explore some new options that may be here sooner than you think....

4

Quantum Teleportation: New York to California in 2 seconds, how about it?

Ever since the wheel was invented more than 5,000 years ago, people have been inventing new ways to travel faster from one point to another. The chariot, bicycle, automobile, airplane and rocket have all been invented to decrease the amount of time we spend getting to our desired destinations. Yet each of these forms of transportation share the same flaw: They require us to cross a physical distance, which can take anywhere, from minutes to many hours depending on the starting and ending points.

But what if there were a way to get you from your home to the supermarket without having to use your car or from your backyard to the International Space Station without having to board a spacecraft? There are scientists working right now on such a method of travel, combining properties of telecommunications and transportation to achieve a system called **teleportation**. In Section 4 we will learn about experiments that have actually achieved teleportation with photons, and how we might be able to use teleportation to travel anywhere, at anytime.

What is Teleportation?

Teleportation involves dematerializing an object at one point, and sending the details of that object's precise atomic configuration to another location, where it will be reconstructed. What this means is that time and space could be eliminated from travel -- we could be transported to any location instantly, without actually crossing a physical distance.

Most of us were introduced to the idea of teleportation, and other futuristic technologies, by the short-lived **Star Trek** television series (1966-69) based on tales written by Gene Roddenberry. Viewers watched in amazement as Captain Kirk, Spock, Dr. McCoy and others beamed down to the planets they encountered on their journeys through the universe.

In 1993, the idea of teleportation moved out of the realm of science fiction and into the world of theoretical possibility. It was then that physicist **Charles Bennett** and a team of researchers at **IBM** confirmed that **quantum teleportation** was possible, but only if the original object being teleported was **destroyed**. This revelation, first announced by Bennett at an annual meeting of the American Physical Society in March 1993, was followed by a report on his findings in the March 29, 1993 issue of Physical Review Letters. Since that time, experiments using photons have proven that quantum teleportation is in fact possible.

Photon Experiments

In 1998, physicists at the California Institute of Technology (Caltech), along with two European groups, turned the IBM ideas into reality by successfully teleporting a **photon**, a particle of energy that carries light. The Caltech group was able to read the atomic structure of a photon, send this information across 1 meter (3.28 feet) of coaxial cable and create a **replica** of the photon. As predicted, the original photon no longer existed once the replica was made.

In performing the experiment, the Caltech group was able to get around the **Heisenberg Uncertainty Principle**, the main barrier for teleportation of objects larger than a photon. This principle states that you cannot simultaneously know the location and the speed of a particle. But, if you can't know the position of a particle, then how can you teleport it? In order to teleport a photon without violating the Heisenberg Principle, the Caltech physicists used a phenomenon known as **entanglement**. In entanglement, at least three photons are needed to achieve quantum teleportation:

- Photon A: The photon to be teleported

- Photon B: The transporting photon

- Photon C: The photon that is entangled with photon B

If researchers tried to look too closely at photon A without entanglement, they would bump it, and thereby change it. By entangling photons B and

C, researchers can extract some information about photon A, and the remaining information would be passed on to B by way of entanglement, and then on to photon C. When researchers apply the information from photon A to photon C, they can create an exact replica of photon A. However, photon A no longer exists as it did before the information was sent to photon C.

In other words, when Captain Kirk beams down to an alien planet, an analysis of his atomic structure is passed through the transporter room to his desired location, where a replica of Kirk is created and the original is destroyed.

A more recent teleportation success was achieved at the Australian National University, when researchers successfully teleported a laser beam. While the idea of creating replicas of objects and destroying the originals doesn't sound too inviting for humans, quantum teleportation does hold promise for quantum computing. These experiments with photons are important in developing networks that can distribute quantum information. Professor **Samuel Braunstein**, of the University of Wales, Bangor, called such a network a "quantum Internet." This technology may be used one day to build a quantum computer that has data transmission rates many times faster than today's most powerful computers.

Human Teleportation

We are years away from the development of a teleportation machine like the transporter room on Star Trek's Enterprise spaceship. The laws of physics may even make it impossible to create a transporter that enables a person to be sent instantaneously to another location, which would require travel at the speed of light.

For a person to be transported, a machine would have to be built that can pinpoint and analyze all of the 10^{28} atoms that make up the human body. That's more than a trillion trillion atoms. This machine would then have to send this information to another location, where the person's body would be reconstructed with exact precision. Molecules couldn't be even a

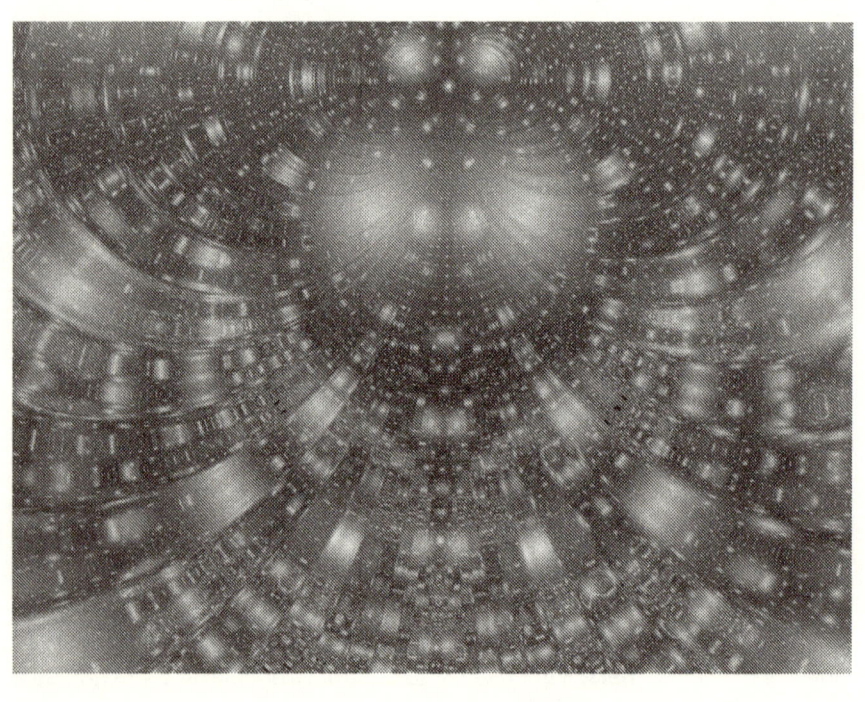

millimeter out of place, lest the person arrive with some severe neurological or physiological defect.

In the Star Trek episodes, and the spin-off series that followed it, teleportation was performed by a machine called a transporter. This was basically a platform that the characters stood on, while Scotty adjusted switches on the transporter room control boards. The transporter machine then locked onto each atom of each person on the platform, and used a transporter carrier wave to transmit those molecules to wherever the crew wanted to go. Viewers watching at home witnessed Captain Kirk and his crew dissolving into a shiny glitter before disappearing, rematerializing instantly on some distant planet.

If such a machine were possible, it's unlikely that the person being transported would actually be "transported." It would work more like a fax machine -- a duplicate of the person would be made at the receiving end, but with much greater precision than a fax machine. But what would happen to the original? One theory suggests that teleportation would combine genetic cloning with digitization.

In this **biodigital cloning**, tele-travelers would have to die, in a sense. Their original mind and body would no longer exist. Instead, their atomic structure would be recreated in another location, and digitization would recreate the travelers' memories, emotions, hopes and dreams. So the travelers would still exist, but they would do so in a new body, of the same atomic structure as the original body, programmed with the same information.

But like all technologies, scientists are sure to continue to improve upon the ideas of teleportation, to the point that we may one day be able to avoid such harsh methods. One day, one of your descendents could finish up a work day at a space office above some far away planet in a galaxy many light years from Earth, tell his or her wristwatch that it's time to beam home for dinner on planet X below and sit down at the dinner table as soon as the words leave his mouth. [19]

So IBM Fellow Charles H. Bennett, confirmed the intuitions of the majority of science fiction writers by showing that perfect teleportation is indeed possible in principle. Other scientists are planning experiments to demonstrate teleportation in microscopic objects, such as single atoms or photons, in the next few years. But science fiction fans will be disappointed to learn that no one expects to be able to teleport people or other macroscopic objects in the foreseeable future, for a variety of engineering reasons, even though it would not violate any fundamental law to do so. Until recently, teleportation was not taken seriously by scientists, because it was thought to violate the uncertainty principle of quantum mechanics, which forbids any measuring or scanning process from extracting all the information in an atom or other object. According to the uncertainty principle, the more accurately an object is scanned, the more it is disturbed by the scanning process, until one reaches a point where the object's original state has been completely disrupted, still without having extracted enough information to make a perfect replica. This sounds like a solid argument against teleportation: if one cannot extract enough information from an object to make a perfect copy, it would seem that a perfect copy cannot be made. But the six scientists found a way to make an end-run around this logic, using a celebrated and paradoxical feature of quantum mechanics known as the Einstein-Podolsky-Rosen effect. In brief, they found a way to scan out part of the information from an object A, which one wishes to teleport, while causing the remaining, unscanned, part of the information to pass, via the Einstein-Podolsky-Rosen effect, into another object C which has never been in contact with A. Later, by applying to C a treatment depending on the scanned-out information, it is possible to maneuver C into exactly the same state as A was in before it was scanned. A itself is no longer in that state, having been thoroughly disrupted by the scanning, so what has been achieved is teleportation, not replication.

As the figure suggests, the unscanned part of the information is conveyed from A to C by an intermediary object B, which interacts first with C and then with A. What? Can it really be correct to say "first with C and then with A"? Surely, in order to convey something from A to C, the delivery vehicle must visit A before C, not the other way around. But there is a subtle, unscannable kind of information that, unlike any material cargo,

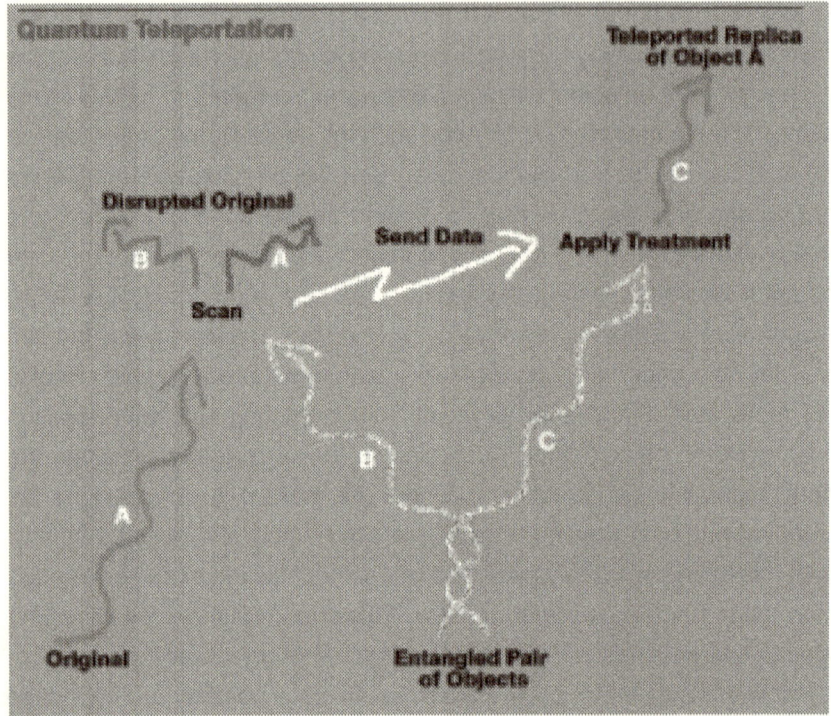

and even unlike ordinary information, can indeed be delivered in such a backward fashion. This subtle kind of information, also called "Einstein-Podolsky-Rosen (EPR) correlation" or "entanglement", has been at least partly understood since the 1930s when it was discussed in a famous paper by Albert Einstein, Boris Podolsky, and Nathan Rosen. In the 1960s John Bell showed that a pair of entangled particles, which were once in contact but later move too far apart to interact directly, can exhibit individually random behavior that is too strongly correlated to be explained by classical statistics. Experiments on photons and other particles have repeatedly confirmed these correlations, thereby providing strong evidence for the validity of quantum mechanics, which neatly explains them. Another well-known fact about EPR correlations is that they cannot by themselves deliver a meaningful and controllable message. It was thought that their only usefulness was in proving the validity of quantum mechanics. But now it is known that, through the phenomenon of quantum teleportation, they can deliver exactly that part of the information in an object which is too delicate to be scanned out and delivered by conventional methods.

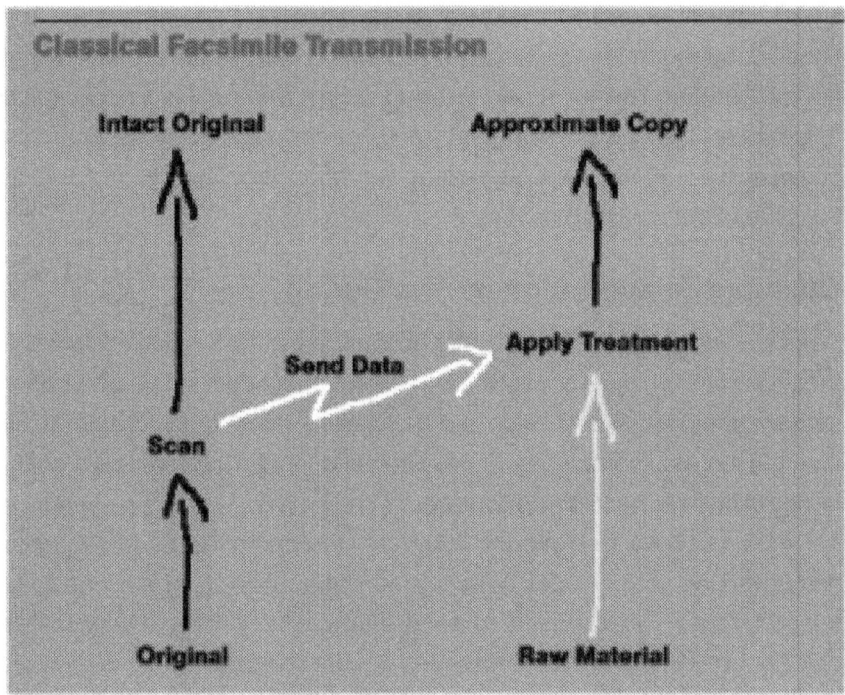

This figure compares conventional facsimile transmission with quantum teleportation (see above). In conventional facsimile transmission the original is scanned, extracting partial information about it, but remains more or less intact after the scanning process. The scanned information is sent to the receiving station, where it is imprinted on some raw material (eg paper) to produce an approximate copy of the original. In quantum teleportation two objects B and C are first brought into contact and then separated. Object B is taken to the sending station, while object C is taken to the receiving station. At the sending station object B is scanned together with the original object A which one wishes to teleport, yielding some information and totally disrupting the state of A and B. The scanned information is sent to the receiving station, where it is used to select one of several treatments to be applied to object C, thereby putting C into an exact replica of the former state of A. [20]

Light Transport

What physicists at Caltech, Aarhus University in Denmark and the University of Wales have accomplished is to take something—a beam of light, in this case—and create a replica some distance away. "We claim this is the first bona fide teleportation," says Caltech physics Professor Jeff Kimble, one of the researchers. The advance won't lead to *Star Trek* technology, but could help with sophisticated cryptography and possibly ultra-powerful "quantum computers." Kimble and his colleagues report their findings in the Oct. 23 issue of the journal *Science.*

Quantum Teleportation in Your Future

What's noise now could eventually become messages. Scientists hope that quantum computers, which move information about in this way rather than by wires and silicon chips, will be infinitely faster and more powerful than present-day computers. "I believe that quantum information is going to be really important for our society," Kimble says. "Not in five years or 10 years, but if we look into the 100-year time frame it's hard to imagine that advanced societies don't use quantum information." And in principle, teleportation could be used to send information to create replicas of objects, not just light beams. Researchers are already looking to teleport atoms. Could this mean the transporters of *Star Trek* could one day be a reality? "I don't think anybody knows the answer," Kimble said. "Let's don't teleport a person—let's teleport the smallest bacterium. How much entanglement would we need to teleport such a thing?" Would such a teleported bacterium actually be the same bacterium, or just a very good copy? "Again," Kimble says, "no one knows for sure." [21]

5

Computer Engineering: Distributed Computing, Quantum Computers, Scalable TCP

No matter how fast computers become, new applications are found that stretch the available resources to the limit. New advances in networking - from mobile wireless to high-speed photonics, are creating broad opportunities for computer engineering.

Recent advances in technology are providing faster microprocessors and network communications, reducing power dissipation in electronic systems, and producing higher-density, low-cost data storage devices. In turn, these advances are creating a demand for new multi-media applications and interfaces. The result will be a combined world-wide network infrastructure that will service television, telephone, and computer communications. It will also allow a new degree of distributed computing - supporting virtual supercomputers and distributed control systems that tie together the fast-growing number of personal computers and micro-controllers.

Broad new areas of future research will be based upon these advances. Computer Engineering will be the driving force for systems that we can now only imagine, such as: tele-robotics; tele-medicine; virtual instruments; and intelligent highways, homes, and vehicles

Distributed Computing

As microprocessor throughput approaches the speed limitations imposed by fundamental device technology, computational parallelism becomes the most viable alternative for achieving breakthroughs in computing power. Just as hardware advances drive multimedia applications, new multi-media applications, in turn, also increase the appetite for more computational power by tightly coupling computers, from microprocessors to complete workstations, near or far, one can share memory, disk, and computational resources already at hand to greatly enhance processing power. Distributed computing provides the infrastructure to bind large numbers of computers, even those connected through wireless base stations, into a single coherent virtual system that meets the challenge of new applications.

Achieving this goal requires research in a number of areas, such as: Network support for arbitration, routing, and fault tolerance; a scalable distributed memory infrastructure to maintain memory coherence, allow code portability, and simplify programming; new synchronization techniques to handle the coordination of independent, high-speed systems and modules; and advanced computer system architecture and tools that efficiently exploit program parallelism.

Photonic Networks and Components

Photonic Networks research deals with both devices and systems. It also addresses issues related both to free-space optical interconnects and to the optical fiber transmission and communication network. The device and component research is an existing strength of the Department while system research is carried out primarily in the communication theory group. A newly funded team research program is the **Focused Research Initiative** (FRI) on Photonic imaging networks. It intends to push the limits of photonic networks for use in high speed transmission of images and image-format data. In our program, special emphasis is placed on applying these photonic imaging networks for the transmission of medical and biomedical images. Research is being done in the four major directions in this FRI are:

1. The investigation of architecture and protocols suitable for ultra high bandwidth photonic networks.

2. The use of ultra short optical pulses and complex signal-temporal optical processing for the transmission of images at terabit per second rate.

3. The investigation of devices and sub-systems for transparent and all-optical networks operating at aggregate rates of one terabit per second.

4. The investigation of efficient interfacing between the photonic and wireless networks to achieve global distribution and access of image and image-format information.

This research in Distributed Computing and Photonic Networks is being carried out by the faculty at the Jacobs School of Engineering in various groups, in collaboration with the Department of Bioengineering, the Departments of Ophthalmology and Neuroscience of the School of Medicine, and the San Diego Supercomputer Center. [22]

How Quantum Computer Will Work

The massive amount of processing power generated by computer manufacturers has not yet been able to quench our thirst for speed and computing capacity. In 1947, American computer engineer **Howard Aiken** said that just six electronic digital computers would satisfy the computing needs of the United States. Others have made similar errant predictions about the amount of computing power that would support our growing technological needs. Of course, Aiken didn't count on the large amounts of data generated by scientific research, the proliferation of personal computers or the emergence of the Internet, which have only fueled our need for more, more and more computing power.

Will we ever have the amount of computing power we need, or want? If, as **Moore's Law** states, the number of transistors on a microprocessor continues to double every 18 months, the year 2020 or 2030 will find the circuits on a microprocessor measured on an atomic scale. And the logical next step will be to create **quantum computers**, which will harness the power of atoms and molecules to perform memory and processing tasks. Quantum computers have the potential to perform certain calculations billions of times faster than any silicon-based computer.

Scientists have already built basic quantum computers that can perform certain calculations; but a practical quantum computer is still years away. As another part of section 5, we will address just what a quantum computer is and what it'll be used for in the next era of computing.

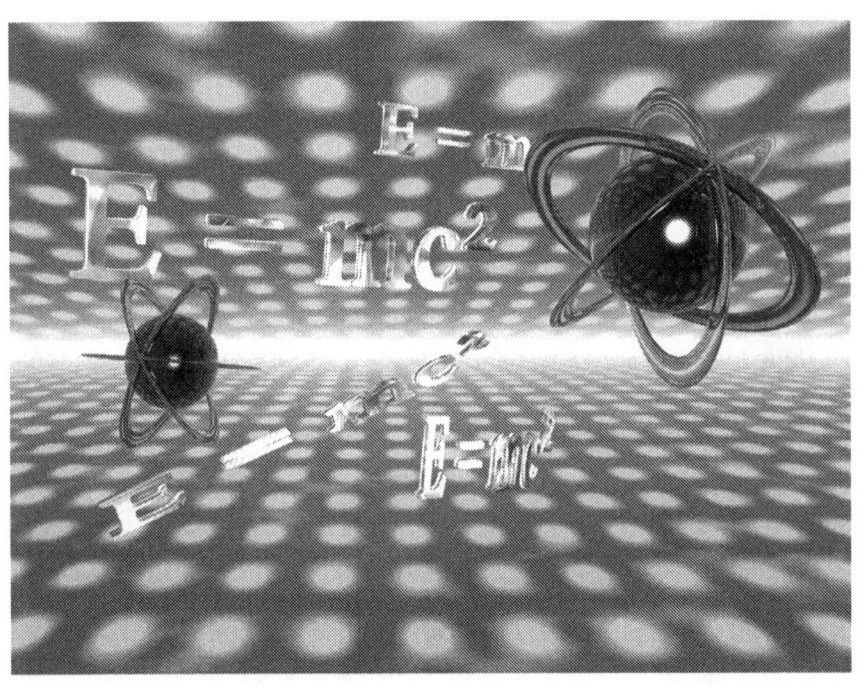

Defining the Quantum Computer

You don't have to go back too far to find the origins of quantum computing. While computers have been around for the majority of the 20th century, quantum computing was first theorized just 20 years ago, by a physicist at the Argonne National Laboratory. **Paul Benioff** is credited with first applying quantum theory to computers in 1981. Benioff theorized about creating a quantum Turing machine. Most digital computers, like the one you are using to read this article, are based on the **Turing Theory**.

The Turing machine, developed by **Alan Turing** in the 1930s, consists of tape of unlimited length that is divided into little squares. Each square can either hold a symbol (1 or 0) or be left blank. A read-write device reads these symbols and blanks, which gives the machine its instructions to perform a certain program. Does this sound familiar? Well, in a quantum Turing machine, the difference is that the tape exists in a quantum state, as does the read-write head. This means that the symbols on the tape can be either 0 or 1 or a **superposition** of 0 and 1. While a normal Turing machine can only perform one calculation at a time, a quantum Turing machine can perform many calculations at once.

Today's computers, like a Turing machine, work by manipulating bits that exist in one of two states: a 0 or a 1. Quantum computers aren't limited to two states; they encode information as quantum bits, or **qubits**. A qubit can be a 1 or a 0, or it can exist in a superposition that is simultaneously both 1 and 0 or somewhere in between. Qubits represent atoms that are working together to act as computer memory and a processor. Because a quantum computer can contain these multiple states simultaneously, it has the potential to be millions of times more powerful than today's most powerful supercomputers.

This superposition of qubits is what gives quantum computers their inherent **parallelism**. According to physicist **David Deutsch**, this parallelism allows a quantum computer to work on a million computations at once, while your desktop PC works on one. A 30-qubit quantum computer would equal the processing power of a conventional computer that could run at 10 **teraflops** (trillions of floating-point operations per second). The fastest supercomputers have achieved speeds of about 2 teraops (trillions of fixed-point operations per second).

Quantum computers also utilize another aspect of quantum mechanics known as **entanglement**. One problem with the idea of quantum computers is that if you try to look at the subatomic particles, you could bump them, and thereby change their value. But in quantum physics, if you apply an outside force to two atoms, it can cause them to become entangled and the second atom can take on the properties of the first atom. So if left alone, an atom will spin in all directions; but the instant it is disturbed it chooses one spin, or one value; and at the same time, the second entangled atom will choose an opposite spin, or value. This allows scientists to know the value of the qubits without actually looking at them, which would collapse them back into 1's or 0's.

Today's Quantum Computers

Quantum computers could one day replace silicon chips, just like the transistor once replaced the vacuum tube. But for now, the technology required to develop such a quantum computer is beyond our reach. Most research in quantum computing is still very theoretical.

The most advanced quantum computers have not gone beyond manipulating more than 7 qubits, meaning that they are still at the "1 + 1" stage. However, the potential remains that quantum computers one day could perform, quickly and easily, calculations that are incredibly time-consuming on conventional computers. Several key advancements have been made in quantum computing in the last few years. Here's a look at a few of the quantum computers that have been developed:

- In August 2000, researchers at IBM-Almaden Research Center developed what they claimed was the most advanced quantum computer developed to date. The 5-qubit quantum computer was designed to allow the nuclei of five fluorine atoms to interact with each other as qubits, be programmed by radio frequency pulses and be detected by nuclear magnetic resonance (NMR) instruments similar to those used in hospitals (see How Magnetic Resonance

Imaging Works for details). Led by Dr. Isaac Chuang, the IBM team was able to solve in one step a mathematical problem that would take conventional computers repeated cycles. The problem, called **order-finding**, involves finding the period of a particular function, a typical aspect of many mathematical problems involved in cryptography.

- In March 2000, scientists at Los Alamos National Laboratory announced the development of a 7-qubit quantum computer within a single drop of liquid. The quantum computer uses NMR to manipulate particles in the atomic nuclei of molecules of trans-crotonic acid, a simple fluid consisting of molecules made up of six hydrogen and four carbon atoms. The NMR is used to apply electromagnetic pulses, which force the particles to line up. These particles in positions parallel or counter to the magnetic field allow the quantum computer to mimic the information-encoding of bits in digital computers.

- In 1998, Los Alamos and MIT researchers managed to spread a single qubit across three nuclear spins in each molecule of a liquid solution of alanine or trichloroethylene molecules. Spreading out the qubit made it harder to corrupt, allowing researchers to use entanglement to study interactions between states as an indirect method for analyzing the quantum information.

If functional quantum computers can be built, they will be valuable in factoring large numbers, and therefore extremely useful for decoding and encoding secret information. If one were to be built today, no information on the Internet would be safe. Our current methods of encryption are simple compared to the complicated methods possible in quantum computers. Quantum computers could also be used to search large databases in a fraction of the time that it would take a conventional computer. But quantum computing is still in its early stages of development, and the technology needed to create a practical quantum computer is years away. Quantum computers must have at least several dozen qubits to be able to solve real-world problems, and thus serve as a viable computing method. [23]

Ultrascale Network Protocols for Computing and Science in the 21st Century

There is a clear and urgent need for multi-Gigabit networks in the scientific community, today, and a need for ultrascale networks in the near future that provide more than 100 Gbps of sustainable throughput end-to-end to transfer Petabyte files. Moreover, continued advances in computing, communication, and storage technologies, combined with the development of national and global Grid systems, hold the promise of providing the required capacities, and an effective environment for the next generation of scientific discoveries.

The **HENP** (High Energy and Nuclear Physics) community has a long tradition of pushing computing and networking technologies to their limits, in production environments. This trend has accelerated in the last few years both due to the Petabytes (1015 bytes) of data acquired, stored, distributed and processed by the worldwide HENP collaborations, and due to the development of Data Grids, which aim to make the data available rapidly and transparently to scientists around the globe. Experiments now underway at SLAC, Fermilab and Brookhaven are already accumulating Petabyte datasets. The next generation of particle physics experiments now under development, due to begin operation in 2007 at CERN in Geneva, will deal with data volumes of tens of Petabytes (in 2007{2008) to Exabytes (1018 bytes) in the decade following. This will impose tremendous new demands on computing, communication and storage technologies.

A current example illustrating the data and computationally intensive character of HENP problems encountered by research teams is the search for Higgs particles at the LHC (the Large Hadron Collider at CERN). A full optimization of the separation of the Higgs discovery signal from potentially overwhelming backgrounds is estimated to require 108 fully simulated and reconstructed background events, drawn from 1011 generated events (sets of simulated four vectors) using loose pre-selection criteria. The processing requirement is approximately 106 CPU-days, or 10,000 of today's fastest processors used round the clock for three to four months. The data resulting from this study will be on the order of 200{400 Terabytes. This implies a need to transfer 2{4 Terabytes per day produced in bursts, which will take

0.5{1 hour of transfer time per day at a throughput of 1 Gbyte/sec end-to-end over the wide area.

Projecting a few years ahead, systems with 10-100 TFlops will be available (the Earth Simulator that was unveiled earlier this year is 36 TFlops) and transfers of 1 Petabyte files will not be uncommon. This requires a global ultrascale network that provides more than 100 Gbps throughput end-to-end. As the decade wears on, this requirement is expected to grow to the Terabit/sec (Tbps) range as the short data \bursts" used by scientists and engineers progress from 1 to 10, and to the 100 Terabyte scale in some cases.

Enabling technologies

The ability to scale silicon technology improves the performance of the devices and decreases their cost, both at an exponential rate. The number of transistors in an integrated circuit has increased by eight orders of magnitude, from two transistors in the 1960s to more than 100 million today. Over the same period, gates have become 1,000 times faster and consume 10,000 times less power. This drastic increase in computing power is more than matched by the advances in communication capacity. The capacity of telephone and data backbone networks has increased by more than three orders of magnitude over the last decade to 10Gbps (the limit of current production-line electronics) on cyber links. The deployment of wave-division multiplexers (WDM) has overcome this electronic speed limit by transporting tens of 10Gbps channels on the same cyber, further increasing the link capacity to several hundred Gbps in the last few years. In the future, it is anticipated that optical cross-connects 2 and other transmission technologies such as solution sources will create pure optical networks where Tbps bandwidth will be common.

Storage technology has also been keeping up, growing from about 2 kbits per square inch in the mid-1950s when disk drives were first invented to 1 Terabit per square inch demonstrated by **IBM** in mid-2002: an increase by nine orders of magnitude.

In parallel to, and driven by, this dramatic increase in performance and decrease in cost of computing, communication, and storage is the spectacular growth of the Internet, from connecting four hosts in 1969 to hundreds of million hosts today, an increase by eight orders of magnitude.

Modeling studies and extrapolations of the rapid advances in computing, communication, and storage technologies show that sufficient capacity will be available for the new generation of scientific computing. The key challenge we face, and intend to overcome, is that our current network control and resource sharing algorithms cannot scale to this regime. Without profound developments in scalable protocols, we cannot build the networks we require to fulfill this vision.

Scalability problems of TCP

A breakthrough that has allowed the Internet to expand by four orders of magnitude in size and five orders of magnitude in backbone speed in the last 15 years was the invention in 1988 by Van Jacobson of the TCP (Transmission Control Protocol) end-to-end flow control algorithm. TCP is a distributed and asynchronous algorithm to share network resources among competing users. It has been carrying more than 90% of the Internet traffic and is instrumental in preventing the Internet from congestion collapse while the Web exploded in the 1990s.

The current TCP however cannot operate efficiently in the bandwidth regime of the future, due to serious equilibrium and stability problems that lead to wildly oscillating transmission rates and even erratic network behavior at high speed. For instance, the current protocol uses packet loss as a congestion measure, which has three difficulties at high speed. First, losses must be extremely rare to support the window size of ultrascale networking. For example, to achieve a throughput of 50 Gbps over a distance with 200 ms round-trip delay using a packet size of 100 kbits will require a loss probability on the order of 10_10. If the current packet size of 12 kbits is used, then the loss probability needs to be on the order of 10_12. This can be difficult to achieve. Second, even if this loss probability is achieved, it is an extremely noisy feedback signal for the sources to reliably use for control.

Finally, since TCP must induce loss in order to estimate the available bandwidth, however rare losses are, when they inevitably occur, they occur in bursts, increasing the likelihood of timeout and underutilization of the network.

Besides the problem with using loss probability for control, the way TCP adapts its rate induces instability at high speed, making wild oscillations unavoidable. These problems must be overcome in order to scale TCP to the multi-Gbps long-distance regime.

TCP's basic flaws, combined with other factors such as limited network access speeds and ill-suited default parameter settings in PCs, have been the culprit behind the observed under-utilization of network resources on major backbones throughout the world. In many countries including the US, these limitations have had a profound economic impact through reduced productivity and lower efficiency of operation of networked systems. As 1-10 Gbps Ethernet, 10 Gbps and faster network backbones, and improved operating systems and default settings are becoming the norm, the main impediment to progress will soon be TCP's lack of scalability.

Scalable protocols

Exciting advances have been made in the last couple years on understanding the equilibrium and dynamic behavior of large networks, such as the Internet. There is now a preliminary theory both to analyze the scalability problems of existing protocols, and to guide the design of new protocols that can in principle scale to arbitrary capacity and delay. We are currently implementing these advances and aim 4 to demonstrate by the end of this year the feasibility of end-to-end flow control in a dynamic environment in the Gbps range over links with 100-200 ms round-trip delay, corresponding to transcontinental and intercontinental distances. Next year w e plan to extend these developments to the 10 Gbps range. The preliminary theory applies to any flow control scheme that works within the decentralization constraints inherent in a large network. It therefore applies not only to TCP, but also to RTP/RTCP or any end-to-end flow control algorithm implemented on top of UDP.

Roderick L. Fennell, M.S. M.I.S.

Despite these advances, much work remains, both on the theoretical and practical fronts. For instance, the current theory is only local and describes the network behavior around equilibrium. A global theory for a distributed network with delay is important. The current implementation effort is restricted to modification of TCP. With changes in active queue management algorithms in routers as well, we will be able to achieve both high utilization and low loss and queuing delay at arbitrary network capacity and size. Other unexplored issues of both theoretical and practical importance include the interaction of control algorithms across protocol layers and across generations in the evolution path.

Conclusion

The development of robust and stable ultrascale networking, at 100 Gbps and higher speeds in the wide area, is critical to support the new generation of ultrascale computing and Petabyte to Exabyte datasets that promise to drive discoveries in fundamental and applied sciences of the next decade. In order to achieve these goals we must, however, reexamine, and, when necessary, redesign, the control protocols that manage these resources, based on a sound and rigorous theoretical foundation, and develop the practical means to scale their capabilities up by orders of magnitude, to meet the demands of future networks and applications.

PART III

RESEARCH METHODS

With the advent of the computer, research that was once only limited to the opening and closing hours of the library can now be accomplished on a 24-hour basis with the invention of the Internet. With that tool at my disposal, it has provided the majority of my research documentation that was presented in this work. I consider that to be an advantage from the old days of college when I attended Texas Tech University for my Undergrad.

To make this explanation as ordered and clear as possible, I will attempt to conical my research by subject matter covered. Outside of the daily influence of my position, I gathered immense enjoyment out of researching my topics from various University sites and Governmental agencies. As to Nanotechnology, I sighted the work of Marvin Minsky of MIT, who is a well known computer scientist and artificial intelligence pioneer. To my amazement also, I found out that IBM had conducted research in the field and that they have been doing so since 1990. Research was conducted as well through the Department of Defense website, NASA, the National Institutes of Health, and the National Institute of Standards and Technology. This research was done as I learned of President Clinton's National Nanotechnology Initiative. The Science Caucus did additional governmental research as to Senate reports by Senator Evan Bayh(D-IN) and Senator Bill Frist(R-TN).

Further reading was done as to Purdue University and Notre Dame Universities work within the field. As to the nanotechnology field, I was also able to find out more about Seminars on the subject as the one to be held in Tokyo on February 2007 called **Nano Tech 2007**. The more I read, the more my ideas were validated as to this emerging Technology.

My research as to Artificial Intelligence consisted as well with MIT, Carnegie Mellon, Stanford, and Cal-Tech researchers. I learned more about AI from excerpts of writings by famed mathematician Alan Turing in his book called, 'Computer Machinery and Intelligence.' Since the term, Artificial Intelligence, was first coined by John McCarthy in 1950, his excerpts were equally valuable. Writer Ray Kurzweil, provided insight also into the fields of Robotics, Genetic Engineering, and Nanotechnology. His Book, **"The Age of Intelligent Machines"**, which was published by MIT, was very helpful as I studied more in the area. Bill Joy, cofounder and Chief Scientist of Sun Microsystems, and co chair on the Presidents commission on the future of IT research, provided great insight from talks with author Ray Kurzweil and his own article from Wired Online magazine. They both offered thought provoking conversations on AI and our technological future.

My research as to Biotechnology and Genetic Engineering was also conducted, via the Web, and touched on work done by University of Southern California's computer scientist Leonard Adleman. His ideas of using DNA to solve complex mathematical equations were profound. He is referred to by some, as the inventor of DNA computers. Intel's founder, Gordon Moore, was also sighted based on his 1965 prediction that microprocessors would double in complexity every two years. Work was also sighted from the American Academy of Arts and Sciences titled, **"The Social Implications of the New Technologies."**

From the area of Cloning, reading was conducted as to Ian Wilmut from the Roslin Institute, and his technique called Somatic Cell Nuclear Transfer (SCNT) as he tried to create the sheep known as Dolly. The website, How Stuff Works, was also very helpful in this and other areas as well. Excerpts were also taken from the BBC's online site from an article titled, **"Genetically Altered Babies Born".**

Researchers from Harvard's Medical School, MIT, and IBM research were referenced in the area of Quantum Teleportation. California Institute of Technology physicists were sighted as they worked to get around the Heisenberg Uncertainty Principle. As to light transport, the Universities of Caltech, Aarhus University of Denmark, and The University of Wales, were sighted.

In the last area of research dealing with Computer Engineering, Kevin Bonsor was very helpful again as he enlightened me in the area of Quantum Computing. As to the thoughts of Distributed Computing and Photonic Networks, the Jacobs School of Engineering at the University of California must be given credit for these ground breaking advances. I must also give credit to Julian Bunn from the Center for Advanced Computing Research of CalTech University and his thoughts of scalable TCP from his Paper titled," **Ultrascale Networking Protocols for Computing and Science in the 21st Century"**.

The information in all of these areas from numerous Universities and Governmental agencies would be enough to fill several books. I feel that I have only touched on the subject matter but, it is my hope that I have given you an indication at least of some of the advances we are making in areas we once only dreamed of.

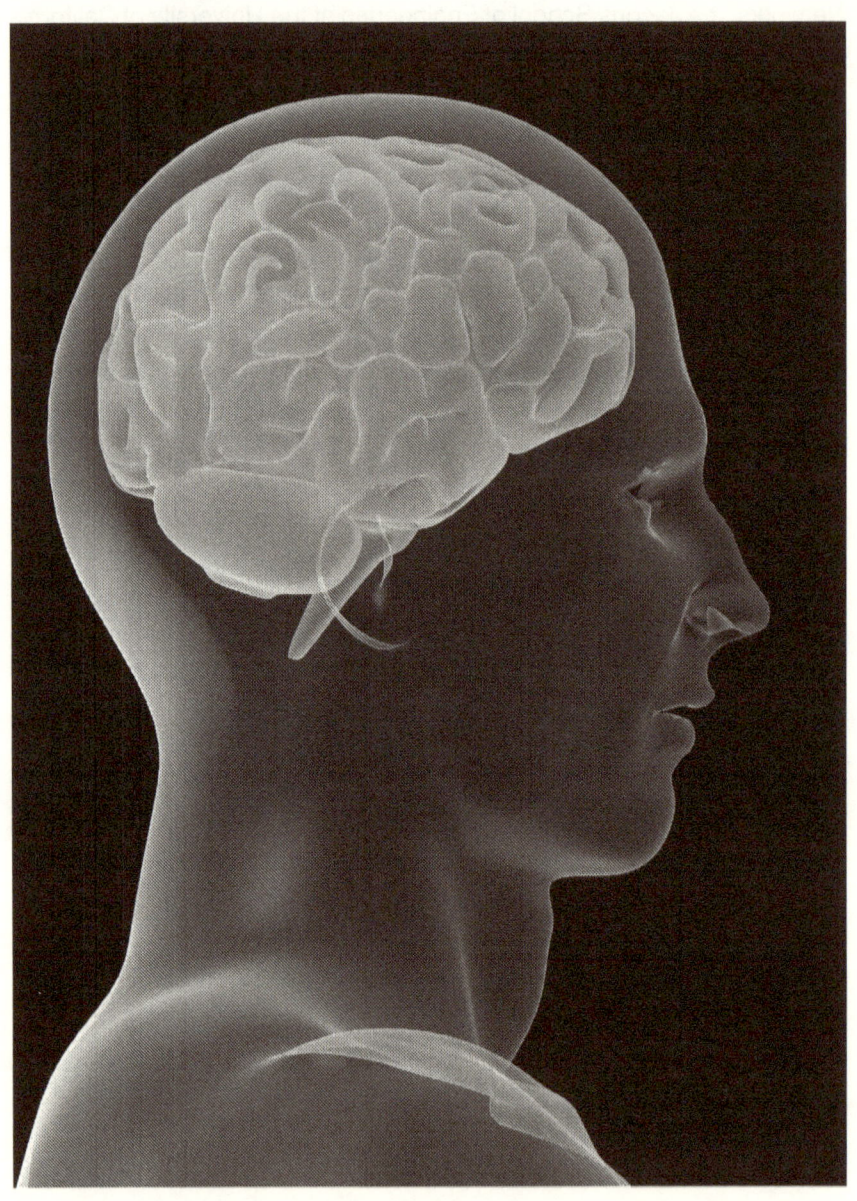

PART IV

OBSERVATIONS AND FINDINGS

The Journey has been long but, well worth the effort. There are so many fascinating breakthroughs and research that are just beginning in these areas that it is very easy to get lost in the writings and still only touch the surface. As to Nanotechnology, I was not aware that there was a Presidential initiative already in place since 2000 and that $227 Million dollars has been set aside for nanotechnology research and development.

Senator Bayh and Senator Frist, I found out, have been involved with this research also and have testified before the Science Caucus committee of the U.S. Senate. It is thought that research in this area will lead to many economic and social advances. I was really surprised to find out that there were conferences being held also along with exhibitions.

As to **AI**, I once thought of this only as science fiction until I read the works from the Massachusetts Institute of Technology and Bell Labs as they worked on artificial neurons that communicated with each other via artificial synapses. This network was composed of transistors fabricated on a silicon integrated circuit to mimic the brains digital and analog synaptic feedback. I learned of the various branches that existed in AI, and some of them were simple games that others or I have played before on the Internet, as in, chess or blackjack. Speech recognition is also an area that I was unaware was a branch of AI.

The thought of having a drink of orange juice and it sending microscopic robots throughout your bloodstream to fight diseases, I must say, is fascinating to me.

The idea of your DNA acting as a supercomputer that can perform complex equations blows my mind. Biochips, which would be non-toxic unlike silicon and cheap, is something to look forward to. The statement that we are getting ready to clone a Human is something I was not aware of as well. I was aware of the Stem research and President Bush's views in this area but, the thought that this could take place outside of our regulations is a distinct possibility. The FDA has prohibited federal funding for human cloning and has the National Bioethics Advisory Commission reviewing the ethical issues associated with this. But, the idea that by 2007, the first human clone may be produced really shocked me.

Genetic research, I was unaware, is being conducted on many areas of our planet. New drugs are being designed to be more effective in fighting diseases such as cancer, diabetes, influenza and hepatitis. I found out that it was also being used to treat dairy cattle, to remove lactose from milk so that people with lactose-intolerance can eat dairy products. It is also being used to cure sickle cell anemia in mice, as well as, being used to provide treatment for enlarged spleens and abnormal urine concentration. So in fact, the ultimate cure for sickle cell anemia may be gene therapy.

As I looked at Star Trek as a child, I never thought that one day Teleportation would ever be possible. I was not aware that Charles Bennett and a team from IBM had confirmed that quantum teleportation was possible by successfully teleporting a photon across 1 meter of coaxial cable and creating a replica of the same photon on the other side. I learned also that the Australian National University researchers were successful in teleporting a laser beam. These experiments will someday aid in the development of networks that can distribute quantum information called, Quantum Computers.

I found out also that there was a newly funded team research program called, **Focused Research Initiative on Photonic imaging networks** that intends to push the limits of photonic networks for use in high speed transmission of images and image-format data. Moore's Law, which states that the number of transistors on a microprocessor continues to double every 18 months and that by the year 2020 or 2030, we will find microprocessor circuits that will be measured on the atomic scale, is one thing more I was not aware of. One other fascinating thing is that Quantum computers will not function the same as today's computers that manipulate bits in two states: a 0 or a 1, but that 3 states will exist. They will have the ability to exist in a superposition that is simultaneous both 1 and 0 or somewhere in between. This is called **qubits**.

Finally, as to Network protocols, I found that the (HENP) High Energy and Nuclear Physics community, that have a tradition of pushing computing and networking technologies to their limits, will be conducting next generation particle physics experiments in 2007 at CERN in Geneva. This research will deal with data volumes of tens of Petabytes (1015 bytes) to Exabytes (1018 bytes) in the decade to follow and will impose high demands on computing, communication, and storage technologies. As to storage technology, I learned that IBM demonstrated that they could store a Terabit of information within 1 square inch.

Since learning is fundamental, I have much more to absorb but I have enjoyed this ride so far. With continued research and developments in these areas, we may all be amazed at what the future may have in store.

PART V

CONCLUSION

Technological advances and trends of the 21st century and beyond will bring about great change for man and mankind. Not only will we have to deal with the economic and governmental changes that will be imposed, we will also have to deal with the social implications of our decisions. Neither you nor I will live to see these advances come to fruition but, it may be issues that our children's children have to face. I do not see that the creation of Nano Machines will be able to happen for at least 100 years or more but research in this area could possibility prove me wrong.

We are already moving in the direction of AI and more advances are to come. But, we must really consider how far we are willing to rely on technology to bring us all the comforts of home. As we work to make the machine more human, we must be careful not to allow human decision making to become a thing of the past. If we do, the ideas of Ray Kurzweil may someday truly happen.

Advances in technology can be an aid to mankind's existence but, if we decide to use cloning to just create a steady supply of human parts with a disregard for the ethical implications of our decisions, we may all lose in the process. It is also true that genetic research may bring about the end of hunger and cure all the diseases we once thought of as incurable but, care must be taken to not use it to create a new race of super humans or for world domination.

Will we ever teleport a human in my lifetime, I personally think not. I believe that we will be able to soon teleport bacteria or small particles but, until more advances are made in this area, I don't believe it will be worth taking the chance.

Quantum computing and advances in networking may come about in the next 100 years. We are already seeing the constant push for more storage space and faster connectivity. This I feel will be consumer driven. We are always trying to do better than the day before and I would not be surprised at what the scientist of our future will be able to create, given the right incentives and funding.

It is my hope that this information will serve as a tool to let others see the possibilities of a focused mind and intellectual collaboration, when our Universities and governmental entities work hand in hand.

So, what was once science fiction is now on the realm of reality. Will humanity and technology truly combine to create a new race of beings, or will we learn as we go and remain the rulers of our own destiny? Only God knows the true answer to this query…

Bibliography

[1] Gayle Permamit and Chris Peterson. 1993. **On the Cutting Edge of Technology**. Prentice Hall Computer Publishing

[2] Kevin Bonsor Copyright 1998-2002 . How Nanotechnology Will Work. Internet http://www.howstuffworks.com/nanotechnology.htm

[3] Kevin Bonsor Copyright 1998-2002 . How Nanotechnology Will Work. Internet http://www.howstuffworks.com/nanotechnology.htm

[4] Evan Bayh; United States Senator, Indiana; News Release; Title: Bayh, First Chair Senate Panel on the Future of Technology; Science Caucus Explores the Potential Impact of Nanotechnology Research; April 5, 2000 http://bayh.senate.gov/www/Press/2000/05APR00pr.htm

[5] Nano Tech 2007; International Nanotechnology Exhibition & Conference; Nano tech executive committee. http://www.nsti.org/Nanotech2006/program.html

[6] Kurzweil, Raymond. 1990. **The Age of Intelligent Machines**. Massachusetts Institute of Technology Publishing.

[7] How Stuff Works. Copyright 1998-2000. Electronic Circuit Replicates Brain Activity. http://www.howstuffworks.com/news-item42.htm

[8] Hans Moravec . *Dec. 1997. When will computer hardware match the human brain? Internet. Robotics Institute. Carnegie Mellon University.* http://www.transhumanlst.com/volume1/moravec.htm

[9] British Broadcasting on the Internet. The Open University. Open2. Net. The Next Big Thing. http://www.open2.net/nextbigthing/ai/ai_in_depth/in_depth.htm

[10] Joy, Bill. Wired Magazine. Apr 2000. Issue 8.04. Why the Future doesn't need us. Internet. http://www.wired.com/wired/archive/8.04/joy.html

[11] American Academy of Arts & Sciences, Martin Malin. November 2000 Newsletter. Social Implications of the New Technologies. http://www. amacad.org/nl11_00/nl1100_3c.htm

[12] Kevin Bonsor Copyright 1998-2002 . How DNA Computers Will Work. Internet http://www.howstuffworks.com/dna-computer.htm

[13] Kevin Bonsor Copyright 1998-2002 . How DNA Computers Will Work. Internet http://www.howstuffworks.com/dna-computer.htm

[14] The United States Food and Drug Administration. July 8, 2002. Use of Cloning Technology to Clone a Human Being. Internet. http://www. fda.gov/cber/genetherapy/clone.htm

[15] The United States Food and Drug Administration. July 8, 2002. Use of Cloning Technology to Clone a Human Being. Internet. http://www. fda.gov/cber/genetherapy/clone.htm

[16] Kevin Bonsor Copyright 1998-2002 . How Human Cloning Will Work. Internet http://www.howstuffworks.com/human-cloning.htm/printable

[17] BBC News Online. May 4, 2001. Genetically Altered Babies Born. Internet. http://news.bbc.co.uk/1/low/sci/tech/1312708.stm

[18] CSIRO Australia. 2000-2002. GeneTechnologyinAustralia. Gene Technology: How is it done? http://genetech.csiro.au/whatisgt.htm

[19] Kevin Bonsor Copyright 1998-2002 . How Teleportation Will Work. Internet http://www.howstuffworks.com/teleportation1.htm

[20] IBM Research. 1995. IBM Corporation. Quantum Teleportation. Internet http://www.research.ibm.com/quantuminfo/teleportation/

[21] *Kenneth Chang* & Reuters. ABC News Internet. 1999. Beam Up the Photons. Internet. http://abcnews.go.com/sections/science/DailyNews/ teleport981022.html

[22]   University of California, San Diego. Jacobs School of Engineering. ECE Department Internet http://www.ece.ucsd.edu/faculty_research/centers/

[23]   Kevin Bonsor. Copyright 1998-2002 . How Quantum Computers Will Work. Internet http://www.howstuffworks.com/quantum-computer.htm/printable

[24]   Bunn, Julian. Center for Advanced Computing Research (CACR), Caltech. Sept.12, 2002. Ultrascale Network Protocols for Computing and Science in the 21st Century. http://netlab.caltech.edu/FAST/

BIOGRAPHY

As a young child, I would go outside at night and gaze at the stars and wonder. My parents, both having their Masters Degree's and beyond, gave me the latitude to question and to dream. As a teenager, I excelled in my studies and was elected by my town to be an exchange student in Europe. I was flown to Washington D.C, where I was briefed by Senators before my flight. I lived with families in Germany, Austria, & England. I also traveled to Holland, France, Switzerland, Liechtenstein, & Czechoslovakia. As an adult, I have also traveled to Italy, Canada, Mexico, & the Bahamas. Meeting people and making friends with individuals from various cultures increased my respect for different beliefs and viewpoints..

I am a graduate of Texas Tech University, with a B.B.A. in Business Management. I also have a Masters Degree in Management Information Systems.

Outside of the various certifications and positions I have earned throughout my life, I realize that I would have nothing if it were not for the ones that struggled before me.

I have over 18 years of management experience in the fields of Information Technology, Insurance, Finance, and Health Care. I currently work as a leader of the Mainframe, Midrange, and Systems Support areas of a major Health Care provider.

Overall, I am still a small town boy that happens to live in the big city. The encouragement of my wife, my mother, and the love of my daughter, has been an inspiration to me. Because of their love, your journey into the hidden secrets of our future can now begin. So, remember back when you were a child and when you still dared to dream. Take that leap of faith again today and read more of the dreams that await you, beyond the 21st Century.